Table of Contents

Chapter 1
The Evolution of Compressed Air..... 1

Chapter 2
Force Transmission
Through a Fluid 5

Force Transmitted Through a Solid
Force Transmitted Through a Fluid
Pascal's Law
Applying Pressure
Area of a Circle
Fluid Pressure to Mechanical Force
Fluid Power Cylinder
What Cylinders Consist Of
How Cylinders Work
Mechanical Force Multiplication
Intensifier
What a Simple Intensifier Consists Of
How Intensifiers Work
Movement Sacrificed
Pressure Scales
Gage Pressure Scale
Measuring Atmospheric Pressure
Absolute Pressure Scale
Vacuum Pressure Scale
How Vacuum is Determined
Pressure Gages
Plunger Pressure Gage
How a Plunger Gage Works
Bourdon Tube Pressure Gage
How a Bourdon Tube Gage Works
Vacuum Gage

Chapter 3
Energy Transmission
Using a Pneumatic System.............. 15

Gases
Molecular Energy
Gases Take Any Shape and Occupy Any Volume
Heat Energy
Gas Temperature and Pressure
Air Compression
Air Expansion
Pneumatic Transmission of Energy
Positive Displacement Compressor
How a Positive Displacement Compressor Works
Inefficiency in a Pneumatic System
Heat of Compression
Friction
Changing Fluid Direction
Flow Rate
Free Air vs. Standard Air
Velocity
Critical Velocity
Pneumatic System Design

Chapter 4
Control of Pneumatic Energy.......... 27

Valves
Control of Pressure
Control of Pressure After a Compressor
Pressure Switch
How a Pressure Switch Works
Safety Relief Valve
How a Safety Relief Valve Works
Pressure Regulator
How a Pressure Regulator Works
Control of Actuator Direction
Double-Acting Cylinder
Directional Control Valve
How a Directional Control Valve Works
Control of Flow Rate
Needle Valves
How a Needle Valve works
A Simple Pneumatic System
Pneumatic Symbols

Chapter 5
Compressors 35

Compressor Types
Positive Displacement Compressors Piston Type Compressors
How a Reciprocating Piston Compressor Works
Vane Compressors
What a Vane Compressor Consists Of
How a Vane Compressor Works
Helical Compressors
What a Helical Compressor Consists Of
Dry Helical Compressors
Oil Flooded Compressor
Single Screw Compressors
Lobed-Rotor Compressors
Centrifugal Compressors
How a Centrifugal Compressor Works
Radial Compressors (Centrifugal)
Axial Compressor (Centrifugal)
Multi-stage Compressors
Two-stage Piston Compressors
How a Two-stage Piston Compressor Works
Capacity Control
Unloading Methods or Output Control
Effect of Altitude on Performance
Effect of Altitude on Electric Motors
Noise
Acceptable Sound Levels
Pinpointing Noise in the Compressor Room
Compressor House Ventilation
Selecting a Compressor for a System
Installation of a Compressor

Chapter 6
Aftercoolers, Driers, Receivers – Air Distribution Systems.................. 49

Compressor Air
Aftercoolers
How an Aftercooler Condenses Water Vapor
Overcompression
Refrigeration
Absorption Process
Adsorption
Receiver Tank
Sizing a Receiver Tank
Piping System
Grid Systems
Unit or Decentralized Systems
Loop Systems
Installation Consideration
High Cost of Air Leaks
Check for Leaks

Chapter 7
Check Valve, Cylinders and Motors 59

How a Check Valve Works
Cylinders
Seals
Stroke Adjusters
Cylinder Mounting Styles
Mechanical Motions
Types of Cylinder Loads
Common Types of Cylinders
Sizing a Cylinder
Stop Tube
Selecting a Stop Tube
Buckling
Sizing a Rod to Prevent Buckling
Cushions
Selecting a Cylinder to Cushion a Moving Load
Flow Rate Into an Air Cylinder
Practical Cylinder Sizing
Pneumatic Tools (Percussive & Rotary)
Percussive Tools
Rotary Motors
Piston Motor
Vane Motors
Turbine Motor
Selecting an Air Motor

Chapter 8
Directional Control Valves 81
Functional Types of Valves
2-way Valve
3-way Directional Valve
3-way Valves in a Circuit
4-way Directional Valve
4-way Valves in a Circuit
Spring Offset
Normally Open and Normally Closed Valves
Detents
Three-position Valves
Center Condition
Both Cylinder Ports Open to Pressure
Both Cylinder Ports Open to Exhaust-
Pressure Blocked
All Ports Blocked
Basic Valve Design
Shear Action Valves
Characteristics of Sliding Plate Valves
Considerations with Sliding Plate Valves
Packed Spool
Characteristics of Packed Spool Valves
Considerations with Packed Spool Valves
Packed Bore
Characteristics of Packed Bore Valves
Considerations with Packed Bore Type Valves
Packed Spool or Packed Bore
Lapped Spool Valves
Considerations with Lapped Spool Valves
Poppet Valves
Characteristics of a Poppet Valve
Considerations with a Poppet Valve
Directional Valve Actuators
Solenoid Operation
Direct-Acting Solenoid Valves
Pilot Actuated Valves
Internally Piloted Valves
Externally Piloted Valves
Solenoid-Controlled/Pilot Operated Valves
Sub-base Mounting
Flow Coefficient - Cv
Sizing of a Valve for Flow (U.S. Units)
Sizing a Valve for Flow (S.I. Units)
Valve Response Time

Chapter 9
Flow Control Valves,
Silencers, Quick Exhausts............. 103
Orifice
Orifice Size Affects Flow
Fixed Orifice
Adjustable Orifice
Ball Valve
Globe Valve
Needle Valve
Flow Control Valves
"Sandwich" Flow Controls
Accuracy of a Flow Control
Placement of a Flow Control in a
Single-Acting Application
Controlling the Speed of a Double-Acting
Cylinder Employing a 4-ported, 4-way Valve
Controlling the Speed of a Double-Acting
Cylinder Employing a 5-ported, 4-way Valve
Multiple Speed Controls
Pneumatic Flow Control Problems
Quick Exhaust Valves
How a Quick Exhaust Valve Works
A Quick Exhaust Valve in a Circuit
Quick Exhaust Used as Shuttles
Noise of Expanding Air
Silencers (Mufflers)
Selecting a Silencer

Chapter 10
Regulators, Excess Flow Valves, Boosters and Sequence Valves 115

Sequence Valve
Sequence Valve in a Circuit
Pressure Regulator
Venting Type Regulator
What the Venting Type Regulator Consists Of
How a Venting Type Regulator Works
Diaphragm Regulator
Pilot Controlled Regulator
What a Pilot Controlled Regulator Consists Of
How a Pilot Controlled Regulator Works
Sizing a Regulator for Flow
A Regulator in a Circuit
Differential Pressure Circuit
Dual Pressure Circuit
Boosters
Air-Oil Tanks
Air-to-Air Booster Circuits
Air-to-Oil Booster Circuits
Excess Flow Valves

Chapter 11
Air Preparation................................. 129

Contaminants in a Pneumatic System
Contaminant Type
The Micrometer Scale
Limit of Visibility
Air Line Filter
What an Air Line Filter Consists Of
How an Air Line Filter Works
Filter Elements
Depth Type Elements
Edge Type Element
Pore Size Air Filter Elements
Normal Rating
Filter Ratings in Practice
Oil Removal Filter
How an Oil Removal Filter Works
Draining a Filter
Automatic Drain
How Automatic Drain Works
Drain in Open Position
Metal Bowls and Bowl Guards
Selecting a Filter
Lubrication
Pulse Type Lubricators
Standard Air Line Mist Lubricator
How a Standard Air Line Mist Lubricator Works
Filling the Standard Air Line Mist Lubricator
Recirculating Type Lubricator
How a Recirculating Type Lubricator Works
Sizing a Lubricator
The Lubricator in a System
FRL Units
Filters and Lubricators Must be Maintained
FRL Design Considerations

Chapter 1
The Evolution of Compressed Air

Man has used compressed air to transmit useful energy ever since early hunters developed the blowgun. Using their lungs, with capacities of about 6000 in^3/min., they could develop pressures to 1 - 3 psi. Man's lungs were a poor compressor for creating a large useful working force.

Figure 1-1 Primitive man using a blowgun

The development of more powerful and efficient mechanical compressors will provide a good route for tracing man's development of pneumatics as an energy form.

The route, however, has a rather slow start with some notable stumbles.

To start, the first compressors, probably simple bellow-like devices, were developed sometime prior to 3000 BC to provide small puffs of air to aid in fire starting. These evolved into larger, but not significantly more sophisticated, units used in basic metal smelting about 1500 BC.

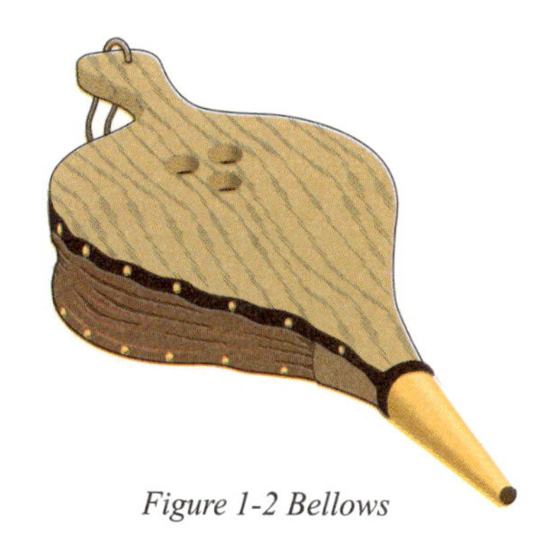
Figure 1-2 Bellows

The development of pneumatics remained relatively static until late in the 18th century when mechanical compressors achieved the capability of generating pressures as high as 1 bar (15 psi).

It was not until the 1800s that compressed air was seriously considered as an industrial energy transfer medium. During this period there were several large scale attempts to use the emerging technology productively.

One of the most notable, if unsuccessful, experiments was the attempt to power a mill with compressed air from a compressor located at a waterfall 3,000 feet from the plant site. The emerging, and not fully understood, technology stumbled here. Clay pipe, fine for transporting water but not air tight, was used to connect the compressor to the remote plant. Air was delivered, but due to clay's permeability, sufficient pressure was not available to turn the mill.

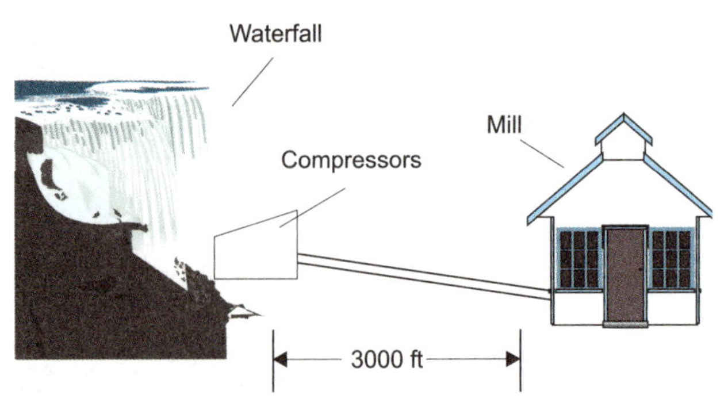

Figure 1-3 The first attempt at using compressed air to power a mill was not successful.

Early in the 19th century, trusting to the potential of compressed air, and assisted by the development of compressors capable of producing 6 bar (90 psi) or about six times the power of early 16th century compressors) pneumatics was selected to power a monumental tunneling project in Mt. Cenis, in the Alps. If manual drilling methods were to be used, it was estimated that the 13.7 km (8.5 mile) tunneling project would require 30 years to complete.

Using pneumatic rock drills, operating from over 4 miles of air lines, the tunnel was successfully completed in 14 years. It was open to traffic in 1871.

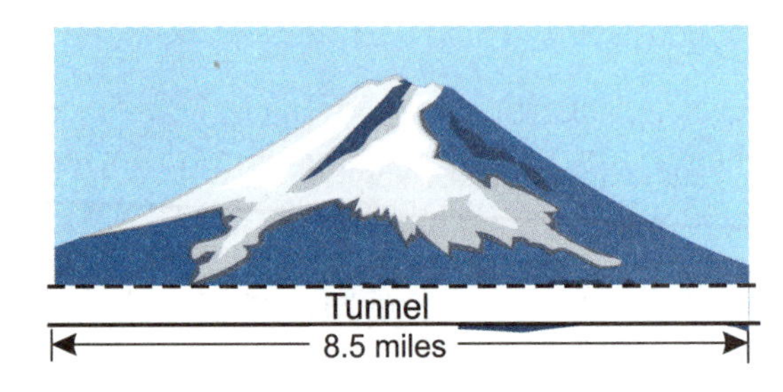

Figure 1-4 Using pneumatic rock drills, an 8.5 mile tunnel was completed in 14 years.

This successful application of the technology attracted international interest, leading many city governments to talk of building central compressor stations for city-wide power.

Paris, the City of Lights, was actually the City of Air. In 1888, Paris installed a 48 kW (65 hp) compressor feeding four miles of mains with 30 miles of branches (a converted sewer system), delivery 6 bar (90 psi) air. By 1891, the capacity was increased to 18642 kW (25000 HP).

Once compressed air was commercially available, pneumatic devices were everywhere; among them were small air-powered electrical generators in restaurants, hospitals and theaters. Admittedly prestige and novelty played a part in the rapid acceptance of pneumatics.

Figure 1-5 Paris, the City of Lights, installed a 48 kW (65 hp) compressor feeding four miles of mains.

Among the many visitors to Paris at the time were engineers, many of whom decided compressed air was the energy transmission system of the future. Conversely, they believed that another emerging technology, electricity, had far too many technical shortcomings to ever be successful.

Obviously, neither extreme was correct. The use of compressed air and electricity expanded during the late 1800s.

In the technological evolution, both electricity and pneumatics each found its place, electricity being the most convenient form for large-scale energy transmission and pneumatics finding specific industrial applications including power and process service and control functions.

The development of mass production on assembly lines as a standard industrial process increased the demand and application possibilities of compressed air. A list of typical industrial applications is presented in Fig. 1-6.

In recent years, compressed air has been applied to control circuitry, dental drills, surgery, as well as to many industrial processes requiring high forces or impact blows. Lightweight, durable and safe pneumatic tools such as pneumatic staplers and pneumatically powered impact wrenches are familiar to many.

Other modern applications of the technology which began with the lung powered blowgun are as varied as compressed air starters for diesel motors, snow making machines for ski slopes and pneumatic lifts in automobile service stations.

This text will introduce you to the basic principles and components found in the typical industrial pneumatic, compressed air system. It is also designed to help you understand basic pneumatic circuits that form an integral part of the "muscle" like parts of the above applications.

Industrial Applications of Compressed Air

air brake	conveying	hoisting
air cylinder	die casting	mixing
air motor	drilling	paint spray
atomizing	elevating	pile driving
buffing	forming	pressurizing
chipping	grinding	process control
reaming	riveting	stapling
screwdriving	transferring	

Figure 1-6 Common uses of compressed air

Chapter 2
Force Transmission Through a Fluid

Pneumatic systems are fluid power systems. They generally use air, a highly compressible, gaseous fluid as the medium of energy transmission. Just as other power transmission systems (mechanical, hydraulic, electrical) pneumatic systems are capable of transmitting a static force (potential energy) as well as kinetic energy. When a force is transmitted through a gas, it happens in a special way. To illustrate this point, we will compare force transmission through a solid and force transmission through a confined liquid (and through gas).

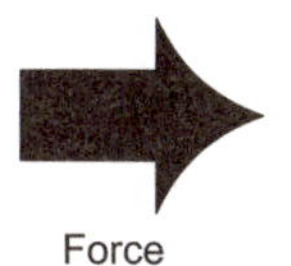

Figure 2-1

Force Transmitted Through a Solid

When a force is applied to a solid, that force is transmitted through the solid and results in a net force acting in the same direction as the original force. As our example at the right shows, if we push on a solid block, the force will be transmitted in the direction of the applied force, through the block to the opposite side.

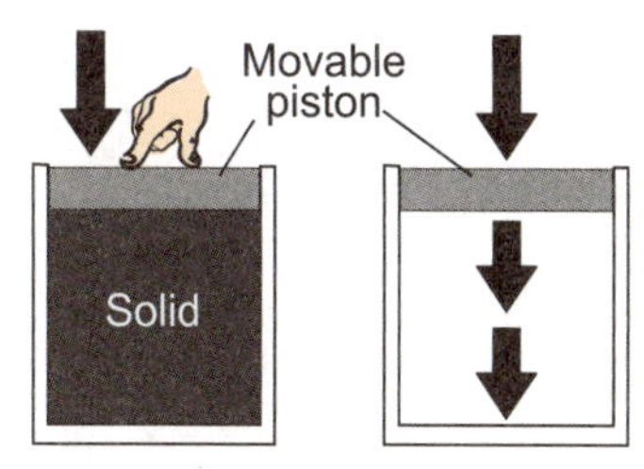

Figure 2-2

Force Transmitted Through a Fluid

But, unlike a solid, a force applied to a confined fluid (gas or liquid) is transmitted equally in all directions throughout the fluid in the form of fluid pressure (generally expressed in psi or bar).

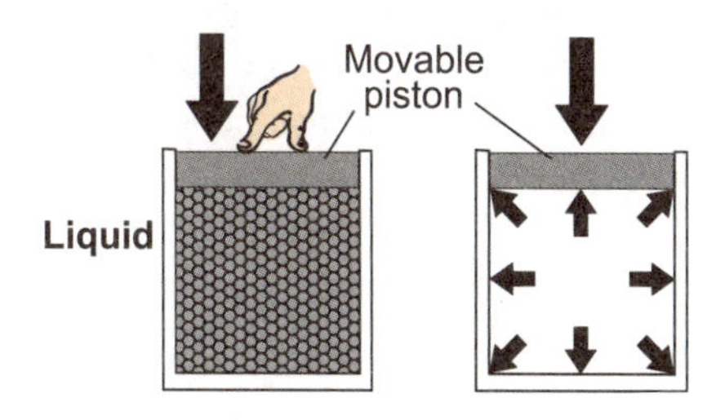

Figure 2-3

To help illustrate this, if the fluid is a liquid (which is not very compressible), the piston will not move appreciably when force is applied. However, in the case of a gaseous fluid (which is generally very compressible), the applied force would push the piston down, compressing the gas. Piston movement would continue until the intensity of the gas pressure, acting against the piston area, equalled the intensity of the applied force.

A confined gas will transmit pressure regardless of how it is generated. As far as the gas is concerned, an applied force will result in pressure whether the application of that force comes from a hammer, by hand, weight, fixed or adjustable spring, or any combination of forces.

All fluids, liquids or gaseous, take the shape of their container. Because of this ability, pressure is transmitted in all directions regardless of the container shape.

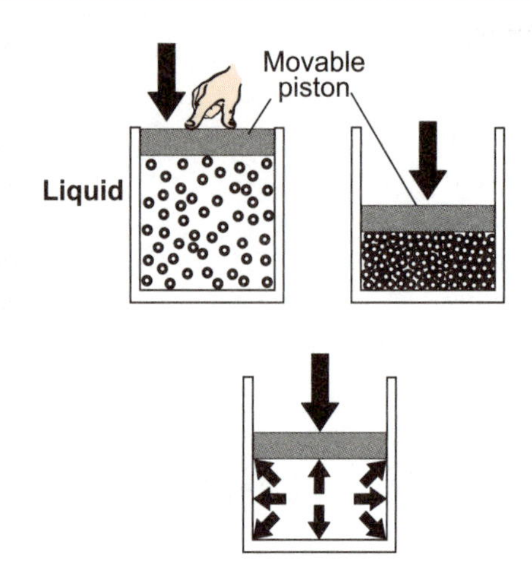

Figure 2-4

Pascal's Law

This property of a gas, to transmit pressure equally in all directions throughout itself, is known as Pascal's Law. This is in honor of Blaise Pascal, who is accepted as having first defined the principle.

Applying Pressure

Thus far, in transmitting pressure through a confined gas, some sort of moveable member has been used to apply the pressure. The most common moveable member is the piston.

Now, in order to determine the intensity of the force being applied to a system, the total force applied must be divided by the effective area of the moveable member, (in this case, the piston). For example, if an applied force of 1000 lbs. (4525N) were applied to a piston with an area of 10 in^2 (65.6 cm^2), the resulting pressure (neglecting seal friction) would be 100 psi (7 bar).

$$\text{pressure} = \frac{\text{force}}{\text{area}} = \frac{1000 \text{ lbs.}}{10 \text{ in}^2} = 100 \text{ psi (7bar)}$$

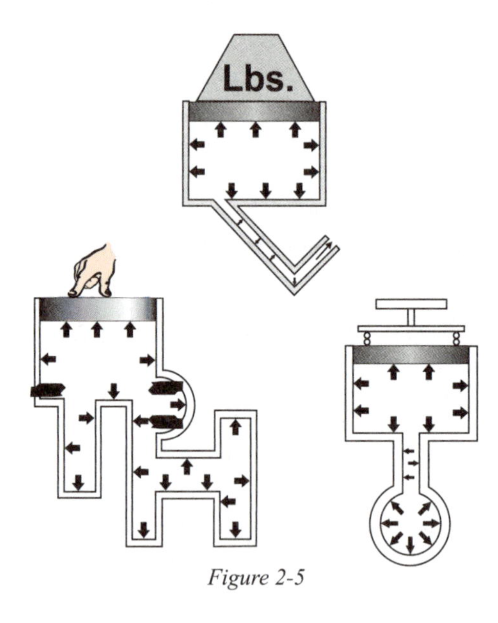

Figure 2-5

Area of a Circle

However, in most cases the area is not known, but the diameter or bore of a piston is. In these cases the piston area must be calculated. The area of a circle (piston) is 78.54% of the area of a square whose sides are the length of the circle's diameter.

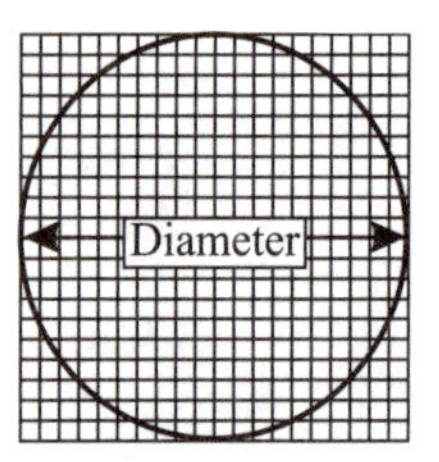

Figure 2-6

To determine the area of a circle, multiply the circle diameter by itself and then by .7854 or by $\frac{\pi}{4}$.

$$\text{circle area} = \text{dia}^2 \text{ x } \frac{\pi}{4} = \text{dia x dia x .7854}$$

There is another way to determine the area of a circle. It is as follows:

$$\text{(radius) x (radius) x 3.14 or } R^2\pi.$$

Fluid Pressure to Mechanical Force

But, applying a force to a gas and transmitting the resulting pressure through the fluid in various shaped containers does not result in useful productive work.

The gas pressure must be converted into mechanical force and motion before useful work can be done. This conversion is the function of a fluid power actuator, that is, to accept gas pressure and convert it into a mechanical force which is capable of moving a "work" piece. The most common type of actuator is the fluid power cylinder.

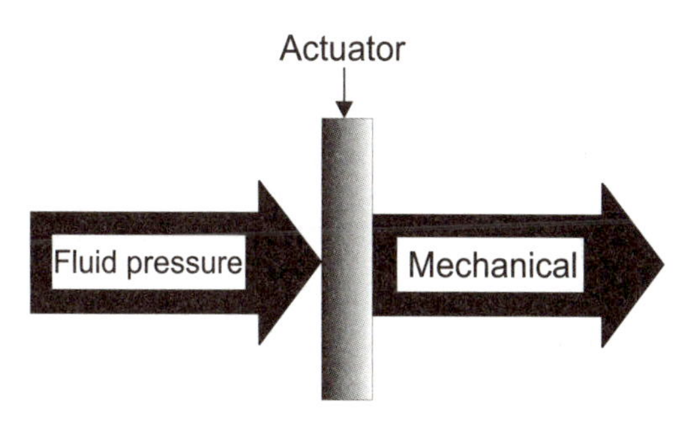

Figure 2-7

Fluid Power Cylinder

The simple fluid power cylinder accepts fluid pressure and converts it into a straight-line, or linear, mechanical force and motion.

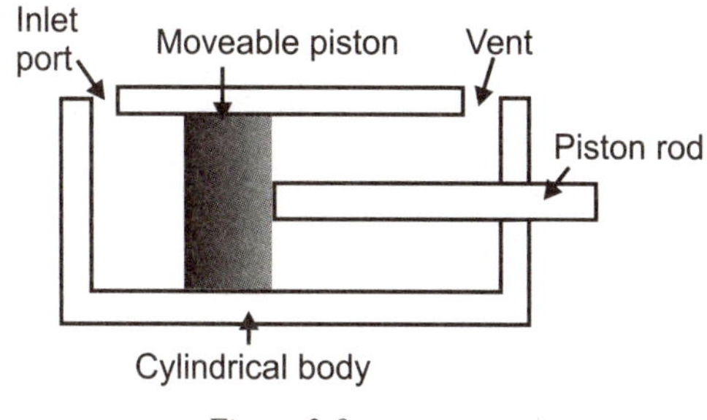

Figure 2-8

What Cylinders Consist Of

A fluid power cylinder basically consists of a cylindrical body, a moveable piston and a rod attached to the piston, a closure with a port at one end, and with a port and an opening for the rod at the other with seals at the piston and rod bearing surfaces. At the "blind" end of an elementary cylinder, the cylinder body has an inlet port by which fluid can enter the body. The other end is open to atmosphere.

How Cylinders Work

With the cylinder inlet port connected to the fluid power system, the cylinder becomes part of the system. In our illustration, when a force is applied by point A, the resulting pressure is transmitted throughout the system and acts on the piston in the cylinder. This results in a mechanical force at point B.

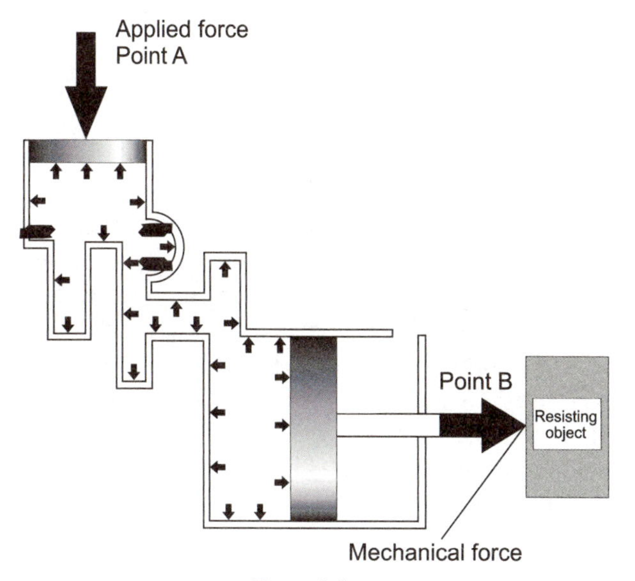

Figure 2-9

Mechanical Force Multiplication

These mechanical forces can be multiplied. The determining factor for force multiplication is the moveable area on which pressure is working. Since pressure is transmitted equally in all directions throughout a confined fluid, if a cylinder piston has more area than the moveable member that is developing the pressure, output force will be greater than input force.

For an example, assume that the resisting object is stationary and will not move. A 1000-lb. (4525N) force on the 10 in^2 (66 cm^2) area piston results in a pressure of 100 psi (7 bar) throughout the system. The 100 psi (7 bar) acts on the cylinder piston with a 15 in^2 (98 cm^2) area, resulting in a mechanical force of 1500 lbs. (6787N).

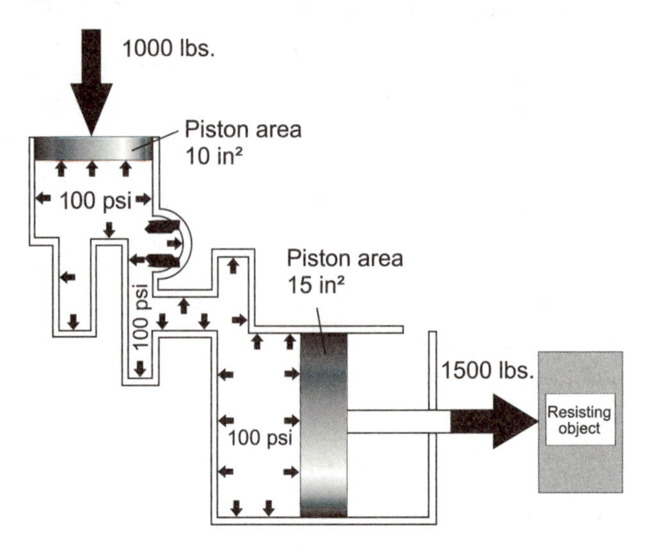

Figure 2-10

For US units:

$$\text{force (lbs)} = \text{pressure} \frac{\text{lbs.}}{\text{in}^2} \times \text{area (in}^2\text{)}$$

For SI units:

$$\text{force (N)} = \text{pressure} \frac{\text{bar.}}{0.1} \times \text{area (cm}^2\text{)}$$

Intensifier

Another method used to increase the fluid actuator force output is to increase the system pressure. One device used to accomplish this is the intensifer. An intensifer multiplies fluid pressure.

What a Simple Intensifier Consists Of

An intensifier basically consists of a housing with inlet and outlet ports, and a large area piston connected to a small area piston. The volume between the two pistons is open to atmosphere (vented).

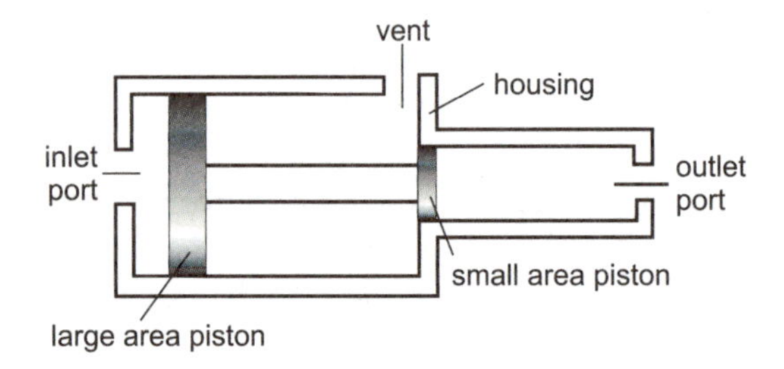

Figure 2-11

How Intensifiers Work

The inlet port of the intensifier is connected to a source of fluid pressure, either air or hydraulic. The intensifier outlet is connected to the part of the system containing fluid where the intensified pressure is desired.

An intensifier will multiply, or intensify, an existing fluid pressure by utilizing a fluid (air or hydraulic) pressure on a large area piston and applying the resultant force to the small area piston. Fluid pressure is therefore intensified or multiplied at the actuator. In our example in Figure 2-13, assume that an object is to be clamped. An input pressure of 100 psi (7 bar) at the intensifier inlet ultimately results in a high output clamping force of 6000 lbs. (27150 N).

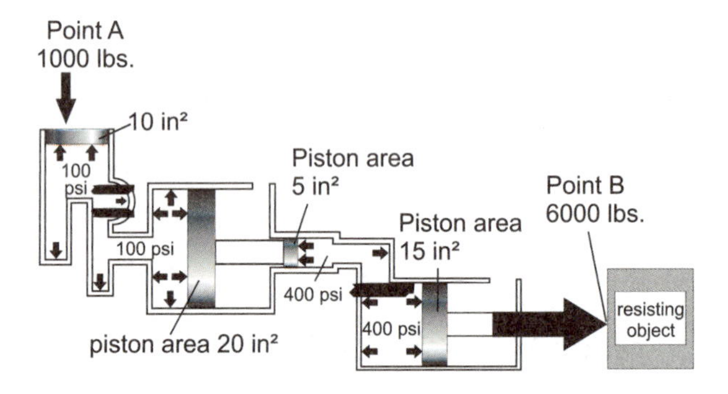

Figure 2-12

Movement Sacrificed

Up to this point, it has been illustrated that a cylinder can be used to multiply a force by the action of fluid pressure acting on a large piston area. When multiplying a force with fluid pressure, it may have appeared that something was received for nothing. It might appear

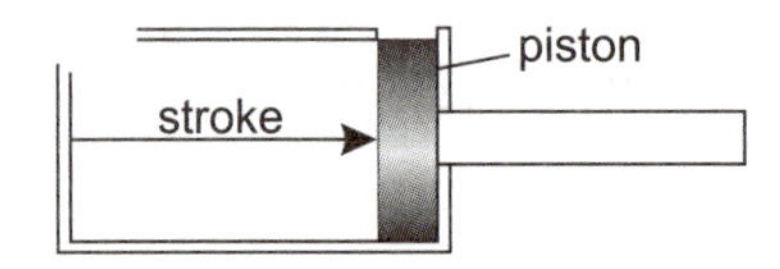

Figure 2-13

that a smaller force could generate a larger force under the right circumstances, and nothing was sacrificed. However, if the force were to be multiplied and moved at the same time, something would be sacrificed. That something is distance. Each cylinder has a stroke and volume. The stroke of a cylinder is the distance through which a piston and piston rod travel.

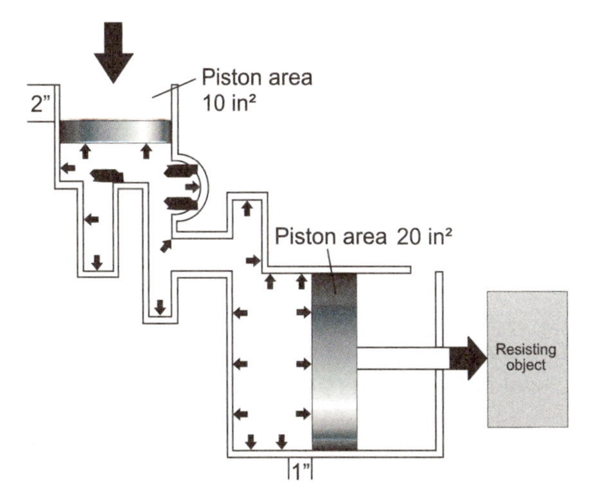

Figure 2-14

The cylindrical volume is the piston's displacement. It is calculated by multiplying the stroke in inches, by the piston area in square inches. This will give the volume in cubic inches.

For US units:

cylinder volume = piston area x stroke

$(in^3) = (in^2) \times (in)$

For SI units:

cylinder volume = piston area x stroke (cm) x $\left(r \frac{mm}{10}\right)$

$(cm^3) = (cm^2) \times (cm)$

In the illustration, the system is filled with a fluid. The top piston must move through a distance of 2" (51 mm) to make the cylinder piston move 10" (25 mm). In both cases the work done is the same. The top piston displaces 20 in³ (328 cm³) of fluid and the lower piston is displaced by 20 in³ (328 cm³) of fluid.

When forces are multiplied with fluid pressure, movement (distance) is sacrificed by the same ratio as the force increase.

Pressure Scales

Either of two pressure scales are used to measure pressure in a fluid power system – an absolute scale or a gage scale.

Gage Pressure Scale

The gage pressure scale measures pressure relative to the ambient atmosphere and thus begins at the point of atmospheric pressure. The unit of measure used for pneumatic fluid power is psi (bar). An ordinary pressure gage used in a fluid power system operates on this scale. The gage scale indicates fluid pressure exerted by the gas in the system. To measure the total pressure of the gas in the system including atmospheric pressure, the absolute pressure scale is used.

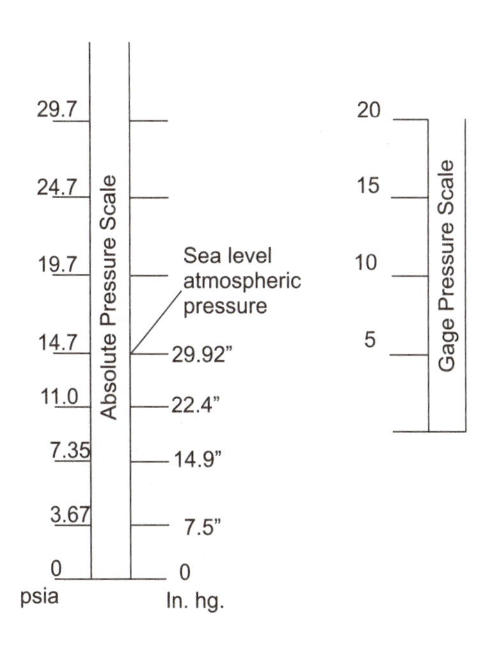

Figure 2-15

Measuring Atmospheric Pressure

Up to this point we have been measuring pressure in pounds per square inch (psi) or (bar). But, it is not uncommon to express low pressures by the height of a liquid column. This is how we measure the pressure exerted by the atmosphere. We generally think of air as being weightless. But, the ocean of air surrounding the earth has weight and thus exerts a pressure.

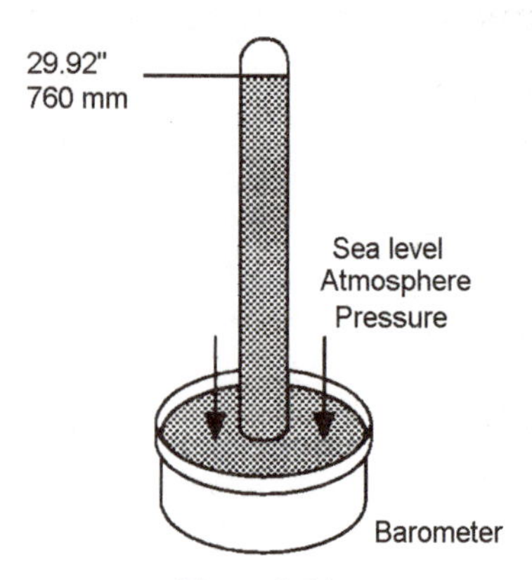

Figure 2-16

Back in the 17th century, a physicist named Torricelli determined that atmospheric pressure could be measured by a column of mercury (Hg). Filling a tube with mercury and inverting it in a pan of mercury, he discovered that on a typical day a standard atmosphere at sea level could support a column of mercury 29.92" high (760 mm Hg). Sea level atmospheric pressure therefore measures, or exerts the same pressure as 29.92" (760 mm) of mercury. Any elevation above sea level would normally measure less. Torricelli's atmospheric measuring device is known as a barometer.

Sometimes it is desirable to convert pressure from inches of mercury to psi. Since one inch of mercury exerts a pressure of 0.491 psi, multiplying a mercury column height by 0.491 converts inches of mercury to psi. For instance, to convert a pressure of 29.92" of mercury to psi, 29.92 is multiplied by 0.491 yields approximately 14.7 psi. Therefore, 29.92" of mercury (standard sea level atmospheric pressure) is reported as a 14.7 psi.

Altitude Above Sea Level in Ft.	Barometer Reading in in. Hg.	Atmospheric Pressure in psi
0	29.92	14.7
1000	28.8	14.2
2000	27.7	13.6
3000	26.7	13.1
4000	25.7	12.6
5000	24.7	12.1
6000	23.8	11.7
7000	22.9	11.2
8000	22.1	10.8
9000	21.2	10.4
10000	20.4	10.0

Mercury column height x 0.491 = psi

Figure 2-17

Absolute Pressure Scale

The absolute pressure scale begins at the point where there is a complete absence of pressure. It can measure and does include atmospheric pressure.

Either psi or inches of mercury are currently common in the U.S. In the SI system the terms mm or Hg are used. To differentiate between the two pressure scales most commonly used in the USA, psig (pounds per square inch gage) is used to denote gage pressure, and psia (pounds per square inch absolute) is used for absolute pressure.

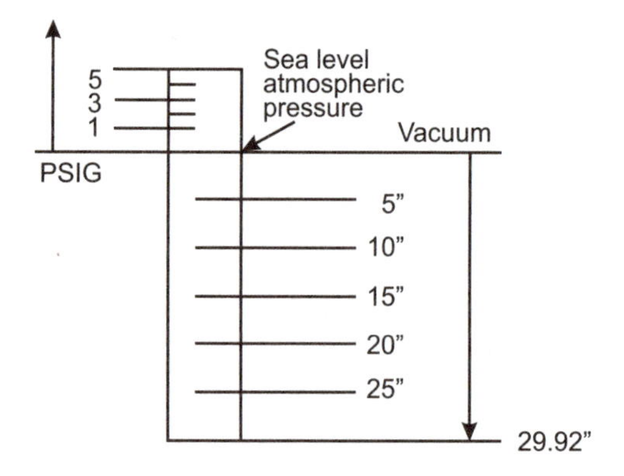

Figure 2-18

Vacuum Pressure Scale

A vacuum is a pressure less than ambient atmospheric. Vacuum pressure is a source of confusion many times because the scale begins at atmospheric pressure, just as gage pressure, but works its way down. In the U.S. it is commonly reported in units of inches (mm) of mercury.

How Vacuum is Determined

In the illustration, a pan of mercury open to the atmosphere is connected by means of a tube to a flask which has the same pressure as the atmosphere. Since the pressure inside the flask is the same as the pressure acting on the pan of mercury, a column of mercury cannot be supported in the tube. Atmospheric pressure is balanced by the pressure in the flask. Zero inches (0.00 mm) of mercury (Hg) in the tube indicates a no-vacuum condition in the flask. If the flask were evacuated so that pressure inside were reduced by 10 inches (254 mm) of mercury, atmospheric pressure acting on the mercury in the pan would support a column of mercury 10 inches (254 mm) high. Atmospheric pressure is balanced by the pressure in the flask plus the pressure exerted by 10 inches (254 mm) mercury. The vacuum would measure 10 (254 mm) Hg. If the flask were evacuated so that no pressure remained and a complete void existed, atmospheric pressure could support a column of mercury 29.92" (760 mm) high at sea level. The vacuum would measure 29.92" (760 mm) Hg. Zero inches of mercury vacuum is atmospheric pressure or the absence of vacuum. 29.92" (760 mm) mercury vacuum indicates near zero absolute pressure of high vacuum at sea level.

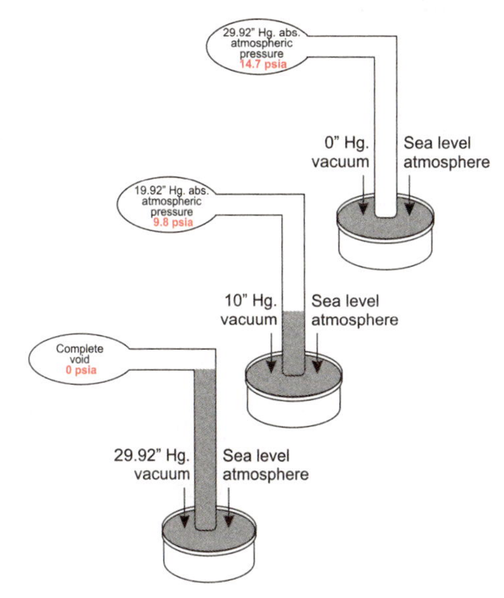

Figure 2-19

Vacuum pressure has an equivalent point on the absolute pressure scale as can be seen when the two scales are compared. For example, at sea level a vacuum of 12" (305 mm) Hg is the same as an absolute pressure of 18" (457 mm) Hg. Sometimes it is helpful to convert vacuum pressure to an absolute pressure.

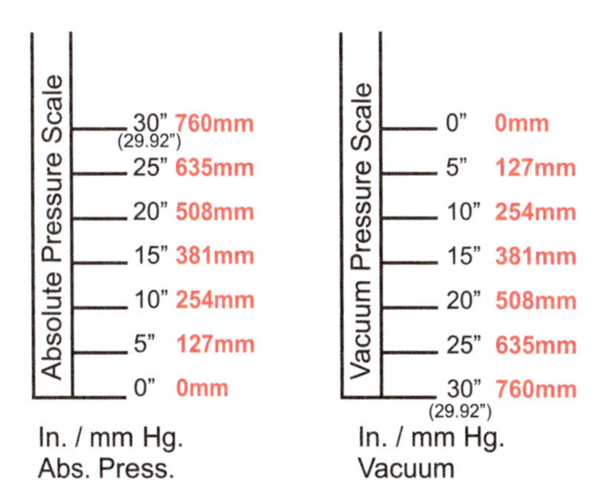

Figure 2-20

Pressure Gages

A pressure gage is a device which measures the intensity of a force applied to a fluid. Two types of pressure gages are most commonly used in a pneumatic system – the Bourdon tube gage and the plunger gage.

Plunger Pressure Gage

A plunger pressure gage consists of a plunger connected to system pressure, a bias spring, pointer and a scale calibrated in pressure units of psi (bar). An example would be auto tire gages.

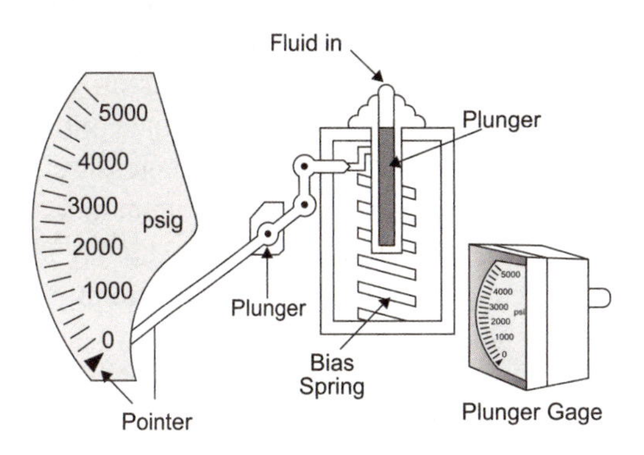

Figure 2-21

How a Plunger Gage Works

As pressure in a system rises, the plunger is moved by the pressure acting against the force of the bias spring. This movement causes the pointer attached to the plunger to indicate the appropriate pressure on the scale. Plunger gages are commonly found in hydraulic fluid power systems. They are a durable and economical means of measuring system pressure.

Bourdon Tube Pressure Gage

A Bourdon tube gage basically consists of a dial face calibrated in the desired pressure units and a needle pointer attached through a Bourdon tube The Bourdon tube is connected to system pressure.

How a Bourdon Tube Gage Works

As pressure in a system rises, the Bourdon tube tends to straighten out. This action causes the pointer to move and indicate the appropriate pressure on the dial face. Bourdon tube gages are relatively precise instruments. They are frequently used for laboratory purposes and on systems where pressure determination is relatively important. Their accuracies range from ±3% to ±.1% of full scale reading, depending upon the accuracy level of the gage. Pressure gages usually measure system pressure which is above atmospheric. The units are in psi (bar) and the scale is gage pressure or psig. To determine an absolute pressure from a gage reading, add the atmospheric pressure to the gage reading. For example, if a machine were operating under normal conditions at sea level and system pressure were 122 psig, the absolute pressure would be 137 psia (122 psig + 14.7 psi [9 bar]).

Vacuum Gage

A vacuum gage is a Bourdon tube gage which measures pressure below atmospheric. A vacuum gage is generally calibrated from 0-30. Each division typically represents the pressure exerted by one inch of mercury. At sea level, to determine an absolute pressure from a vacuum gage reading, subtract the vacuum in inches of mercury from 29.92. For instance, a vacuum reading of 7 in. Hg at sea level is actually an absolute pressure of 22.92" Hg or approximately 23" Hg.

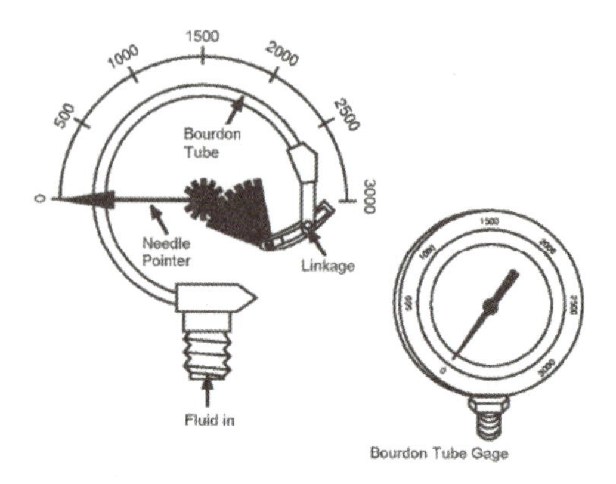

Absolute pressure = Gage reading + 14.7 psi

Figure 2-22

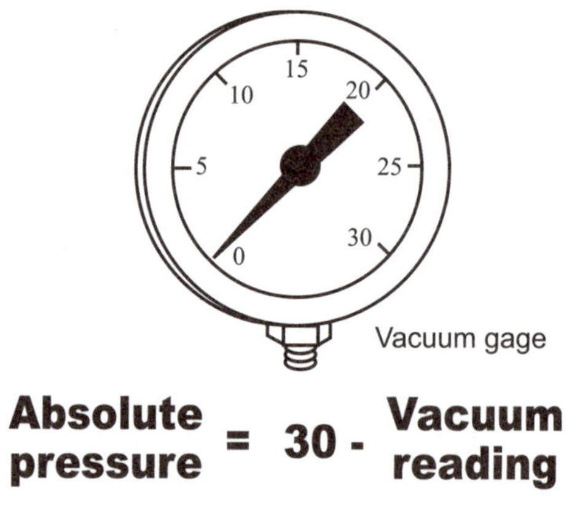

Absolute pressure = 30 - Vacuum reading

Figure 2-23

Lesson Review

In this lesson of force transmission through a fluid we have seen that:

- The intensity of a force applied to a confined liquid or gas is transmitted equally in all directions throughout the fluid in the form of fluid pressure. This is Pascal's Law.
- A fluid power cylinder converts fluid pressure into straight-line, or linear, mechanical force.
- The greater the pressure at a cylinder's piston or the larger the piston area, the greater the mechanical output force.
- An intensifier multiplies fluid pressure, resulting in an increased output force at an actuator coupled to it.
- When forces are multiplied with fluid pressure, movement is sacrificed.
- Two pressure scales are used to measure pressure in a fluid power system – an absolute scale or a gage scale.
- It is not uncommon to express low pressures by the height of a liquid column. Atmospheric pressure is commonly measured by the height of a column of mercury it will support.
- A vacuum is a pressure less than atmospheric. Vacuum may be measured by the height of a mercury column. As an absolute measure it can be expressed as a pressure below atmosphere.
- Two types of pressure gages are commonly used in fluid power systems – the Bourdon tube gage and the plunger gage.
- A typical vacuum gage is a Bourdon tube gage which measures pressures below atmospheric. A vacuum gage is calibrated from 0-30.

Exercise
Force Transmission Through a Fluid
50 Points

Instructions: fill in the blanks with a word from the list at the end of the exercise. Words may be used more than once.

1. The _____ of an applied force is transmitted equally in all directions throughout a confined fluid under the form of fluid pressure.
2. When forces are multiplied with a fluid power system, _____ is sacrificed.
3. An actuator converts fluid pressure into _____ force.
4. At sea level, a vacuum gage reads 6" Hg, which is actually 24" Hg on the _____ scale.
5. A cylinder with an 8" diameter piston and a 3" diameter rod has an effective piston area at the rod side of _____ .
6. A _____ pressure gage is generally more accurate than a plunger type gage.
7. An intensifier multiplies fluid _____ .
8. _____ inches of mercury vacuum is the same as atmospheric pressure.
9. In the intensifier illustrated below, a pressure of _____ at the inlet will result in 4000 psi at the outlet.

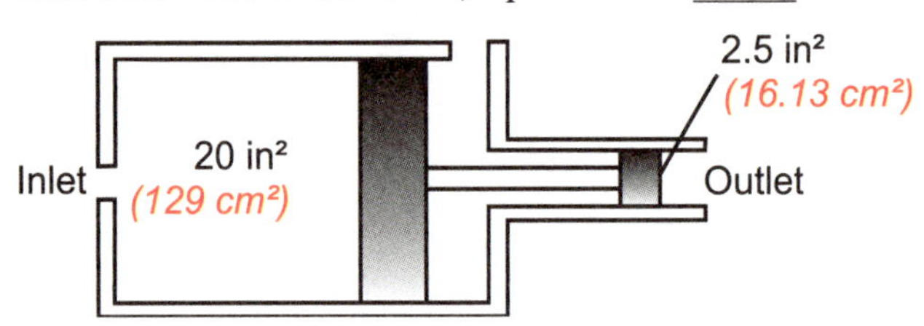

10. In the cylinder illustrated below, a pressure _____ will extend the load.

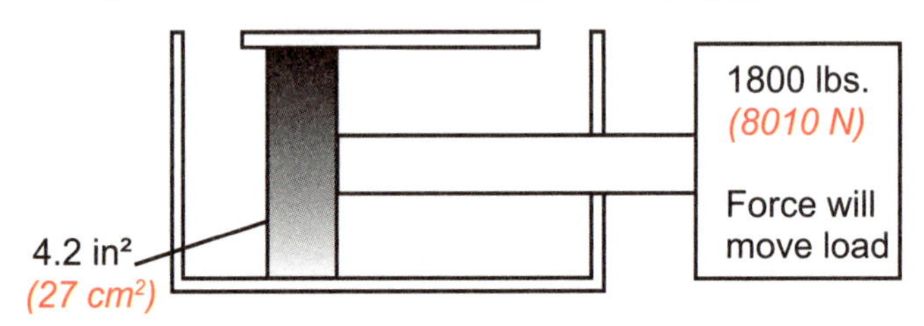

absolute	fluid	intensity	plunger
atmospheric	force	liquid	pressure
barometer	43.2 in²	mechanical	psia
Bourdon tube	400 psi	mercury	psig
converts	gage	movement	stroke
cylinder	gas	19.6 in²	29.92" Hg
diameter²	height	Pascal	vacuum
500 psi	intensifier	piston	zero

Chapter 3
Energy Transmission Using a Pneumatic System

Before dealing with energy transmission through a gas, it will help our understanding of pneumatics to first determine what a gas is and then concentrate on some of its pertinent characteristics.

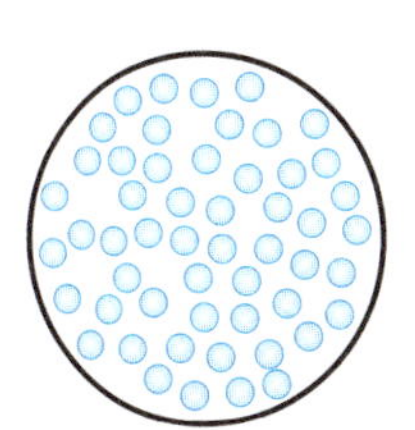

Figure 3-1

Gases

To begin, gas is made up of moving molecules. Unlike molecules of a solid or liquid, gas molecules are not readily attracted to one another, tending to remain separate. Therefore, gas molecules must be housed in a container or they will disperse because of their molecular energy and lack of restraining forces.

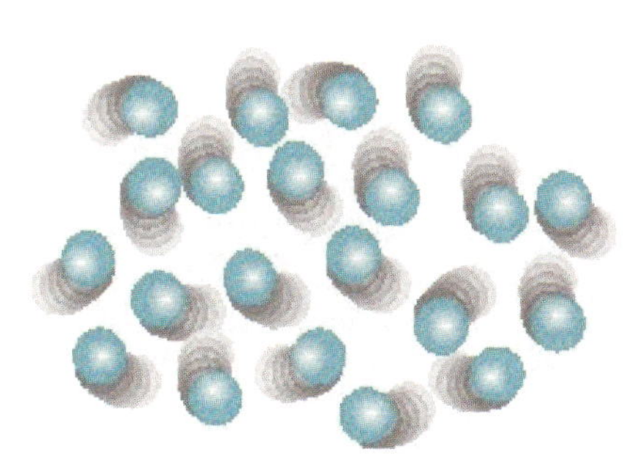

Figure 3-2

Molecular Energy

These molecules which make up a gas do not float around in space like snowflakes. They are continuously speeding throughout their container, crashing into other molecules and the walls of their container. Their movement is more like a swarm of angry bees than a gently falling snow. This molecular energy aids the gas as it takes the shapes and fills the volume of its container.

Gases Take Any Shape and Occupy Any Volume

A large portion of the volume occupied by a gas is space. If space were eliminated, gas molecules would be in continuous contact with one another and would, therefore, be a liquid or solid.

Figure 3-3

With a large portion of space in its makeup and with molecules continuously speeding through that space, a gas can take the shape of any container and fill the volume of that container.

Solids have definite volume and definite shape. Liquids have a definite volume, but conform to the shape of their container. Gases take the shape and volume of their container.

Heat Energy

Now, in order to more fully understand gas properties, we should determine at this point what heat energy is. Heat in a solid is found as vibrating molecules. The hotter an object, the more violently the molecules vibrate.

Liquid molecules are in contact with one another, but are not in a rigid position. They are free to slip and slide past one another. The heat of a liquid is molecules in action. The hotter a liquid is, the quicker the molecular movement.

In a gas, molecules are continuously moving. Heat energy in a gas is this molecular movement. A gas with increased temperature has faster moving molecules than a gas which is cool. We must now examine the relationship between temperature and pressure.

Gas Temperature and Pressure

Under normal conditions, a gas housed in a container has billions of molecules traveling rapidly throughout the volume of the container. The speed of those molecules determines the gas temperature as we saw above.

Each gas molecule, as it speeds through the inner atmosphere of the container, is bombarded by other molecules and crashes into the container walls. Because of the large number of molecules in the container, these collisions occur millions of times per second. If a pressure gage were inserted into the gas container, the gage would interpret the effect of these millions of collisions per second as a single pressure.

With these points in mind, let us see what happens when the volume of a container in which a gas is housed is changed.

First, however, let us examine the relationship of the compressed gas properties of pressure, volume and temperature. The relationship of these properties is expressed by the modified ideal gas law. This law is utilized when two states exist and there is no change in

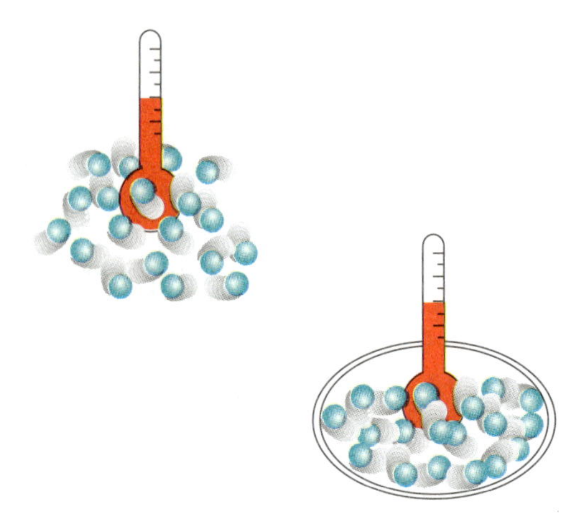

Figure 3-4

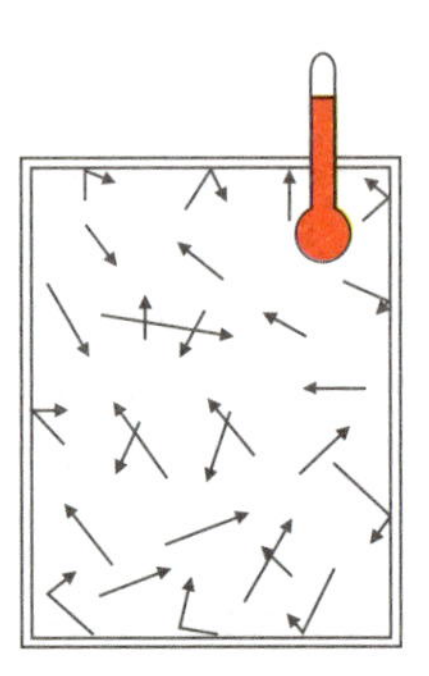

Figure 3-5

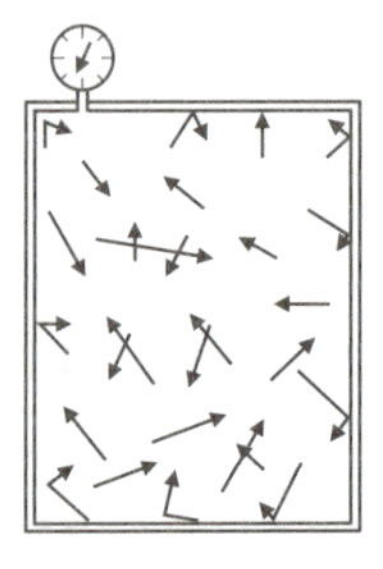

Figure 3-6

the number of molecules of the gas. The mathematical expression of this law is as follows:

$$\frac{P_1V_1}{T_1} = \frac{P_2V_2}{T_2}$$

Where:

P is absolute pressure
T is absolute temperature
V is volume

The subscripts 1 & 2 designate the initial and final state conditions respectively. This equation tells us that at a constant pressure ($P_1 + P_2$) the volume is directly proportional to the absolute temperature. Also, at a constant volume ($V_1 - V_2$) the pressure is directly proportional to the absolute temperature. At a constant temperature ($T_1 = T_2$) the volume will change inversely as the absolute pressure.

This Ideal Gas Law is used when we are interested in single state conditions.

The equation is as follows: $PV = nRT$

Where:

P is absolute pressure
V is total volume
n is number of moles
R is universal gas constant (2271.87 Joules/mole)
T is absolute temperature
Absolute pressure is defined at atmospheric pressure plus gage pressure.
Absolute temperature is Fahrenheit reading plus 460°.

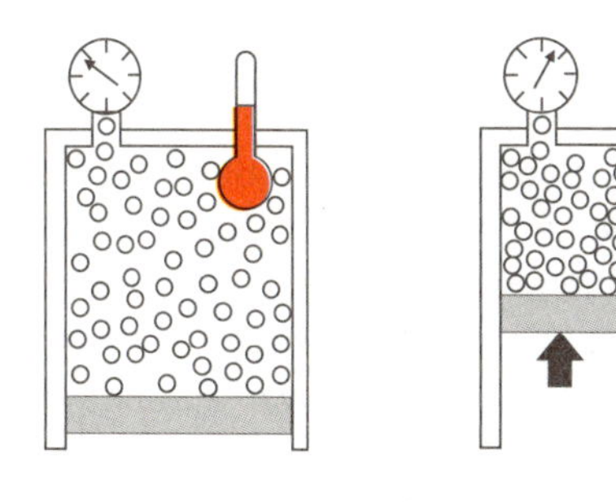

Figure 3-7

This equation will describe the behavior of a real gas at low pressures. Those pressures typically found in today's industrial pneumatic systems.

At these low pressures the accuracy of the Ideal Gas Law is acceptable because of the weak molecular attraction between the molecules.

Air Compression

Now, suppose we had a cylindrical container open to the atmosphere, and assume that the absolute pressure and temperature inside the cylinder are the same as the ambient. (Ambient refers to surrounding conditions.)

If we fitted the cylindrical container with a close fitting piston with a seal, the gas volume would be reduced when the piston was pushed in. While this compressive action was taking place, the number of molecular collisions would increase greatly. This increase is due to the same number of molecules occupying a smaller space. The increase in molecular collisions is reflected in an increase in both temperature and pressure.

After a time, excess molecular energy (heat) would be given up through container walls and gas temperature would once again become equal to the ambient. But since molecules have less of a volume in which to roam, collisions are still more prevalent. Pressure remains at a level higher than atmospheric, but not as high as at the previously elevated temperature.

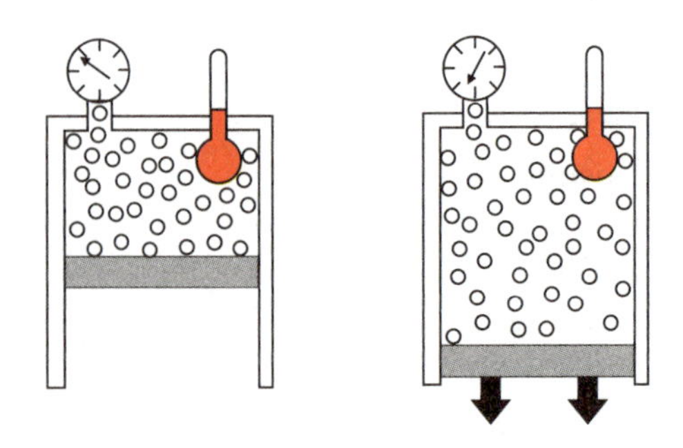

Figure 3-8

When air is compressed, its pressure and temperature increase but what happens if the volume is expanded?

Air Expansion

Suppose we had another cylinder but this cylinder has a close fitting piston midway in its bore.

Let us assume that gas temperature and pressure under the piston are those of the ambient. If the piston were pulled out, gas volume would increase. While expansion was occurring, the number of molecular collisions would decrease greatly. This is because the same number of molecules are occupying a larger volume. This decrease in molecular collisions is reflected in a decrease in both pressure and temperature. Because heat energy would be transmitted through cylinder walls after a time, gas temperature would once again equal the ambient. However, since molecules have more space in which to roam, collisions are still less frequent. Pressure remains below atmospheric, but not as low as when the air was at reduced temperature.

When air is allowed to expand, generally its pressure and temperature decrease.

Pneumatic Transmission of Energy

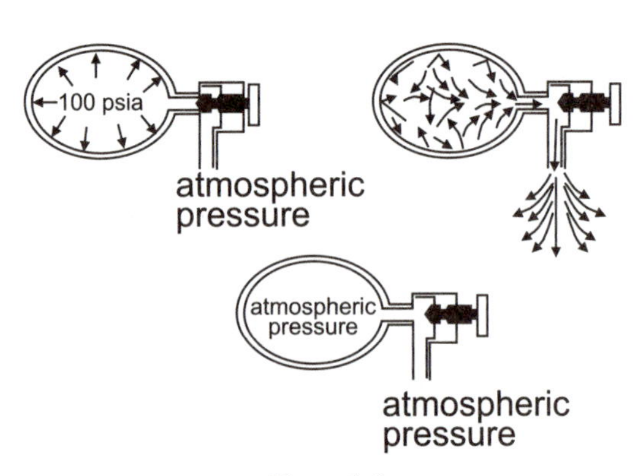

Figure 3-9

The reason for using pneumatics, or any other type of energy transmission on a machine, is to use a convenient source of power to perform work. As illustrated previously, accomplishment of work requires the application of the moving force (kinetic energy) to a resisting object resulting in the object moving through a distance.

In a pneumatic system, energy is stored and distributed in a potential state, in the form of compressed air. Useful work will result from a pneumatic system when the compressed air is allowed to convert its potential energy into kinetic energy. For example, in the illustration a tank is charged to 100 psia (7 bar) with compressed air (potential energy). When the valve at the tank outlet is opened, the air in the tank will flow until the pressure inside the tank equals atmospheric pressure.

To illustrate this point further, two identical tanks are connected by means of a pipe. Midway between the tanks is positioned a shutoff valve. One tank is charged with 100 psia (7 bar) compressed air. The other tank is charge to 50 psia (3.5 bar). If the shutoff valve were opened, the 100 psi (7 bar) air would flow into the other tank until pressures in both tanks were equal. We can use this principle to perform useful work. In another illustration, a tank charged to 100 psia (7 bar) is connected to a cylinder with a shutoff valve separating the two. The cylinder is capable of equaling its load resistance when a pressure of 50 psia (3.5 bar) is present at its piston.

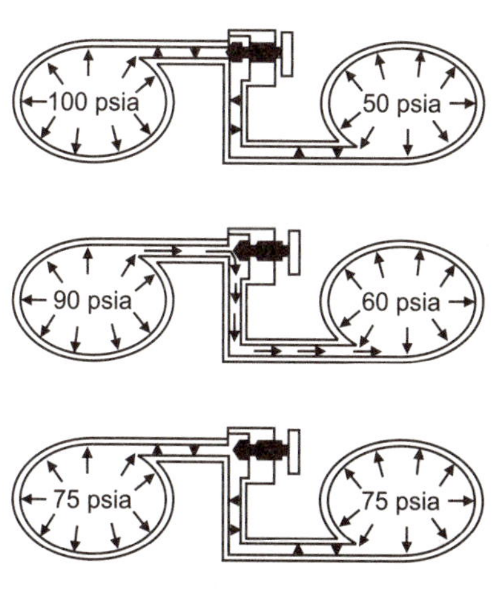

Figure 3-10

When the shutoff valve is opened, the 100 psia (7 bar) tank air will flow into the cylinder. The volume into which the air is flowing remains constant until 50 psia (3.5 bar) is present at the cylinder's piston. At this time, the load resistance is equalled; disregarding friction and other inefficiencies, the cylinder piston and load begin to move; the volume into which air is flowing is growing. Now, work is being done (force moved through a distance). In this type of system, work will continue only as long as enough pressure is present on the piston area to overcome the load resistance. While air is flowing out of the tank, pressure in the tank is dropping. If tank pressure happens to equal cylinder pressure before a cycle is completed, work ceases. Since a tank charged to a pressure can only hold a certain amount of air, a limited amount of work can be done. This is a disadvantage.

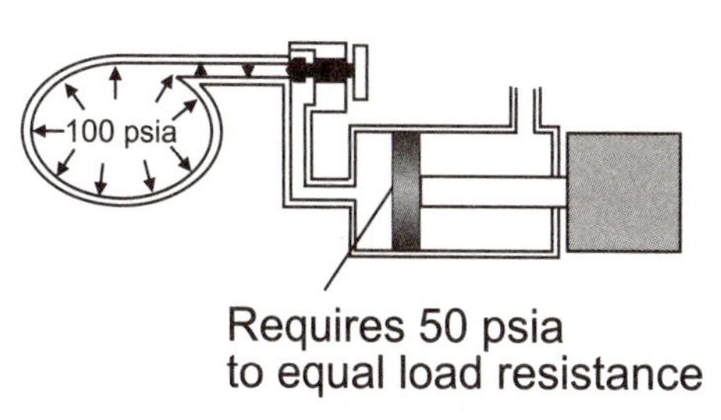

Figure 3-11

To perform any appreciable amount of work then, a device is needed which can supply an air tank with a sufficient amount of air at a desired pressure. This device is an air compressor. One such device is the positive displacement compressor.

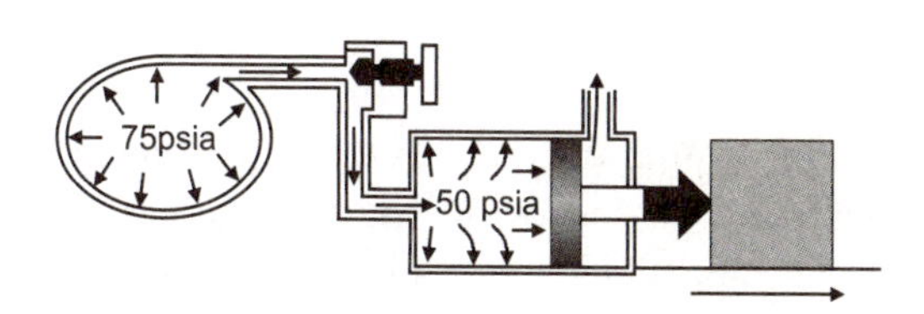

Figure 3-12

Positive Displacement Compressor

A positive displacement compressor delivers compressed air to an air receiver tank. This type of compressor basically consists of a moveable member inside a housing.

The compressor in the illustration has a piston for a moveable member which compresses the air. This piston is driven in a reciprocating manner by a crankshaft which is in turn connected to a prime mover (electric motor, internal combustion engine, etc.). At inlet and outlet ports, valves allow air to enter and exit the chamber.

How a Positive Displacement Compressor Works

As the crankshaft pulls the piston down, an increasing volume is formed within the housing. This action causes the trapped air in the piston bore to expand, reducing its pressure. When pressure differential becomes high enough, the inlet valve opens, allowing atmospheric air to flow in.

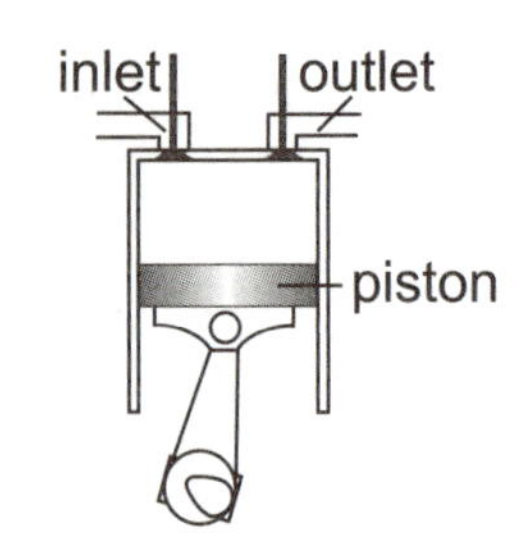

Figure 3-13 Positive displacement compressor components

With the piston at the bottom of its stroke, the inlet valve closes. The piston starts its upward movement to reduce the air volume which consequently increases its pressure and temperature. When pressure differential between the cylinder compressor chamber and discharge line is high enough, the discharge valve opens, allowing air to pass into an air receiver tank for storage and removal of excess moisture and other contaminants. This is our system's potential energy.

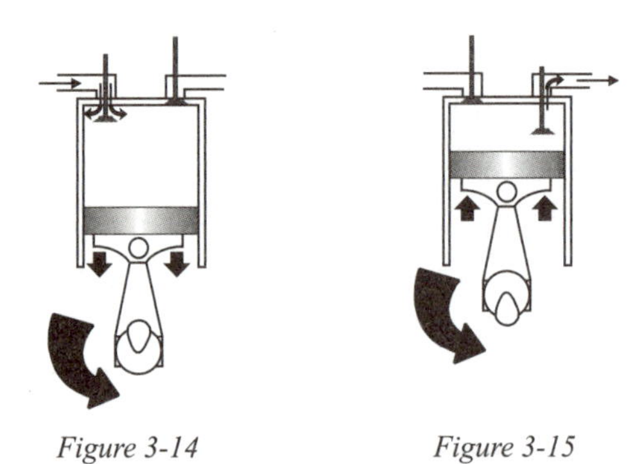
Figure 3-14 *Figure 3-15*

Inefficiency in a Pneumatic System

But, the amount of energy delivered to a pneumatic system is not the same amount of energy delivered to the point of work. This is due to inefficiencies in the system. Some of the pneumatic energy is transformed before it reaches an actuator. The inefficiency in a pneumatic system is generally the result of the necessary loss of heat during air compression, friction and fluid changing direction.

Heat of Compression

Let us look closer at the heat of compression. We saw earlier that when a gas is compressed, its temperature and pressure increase.

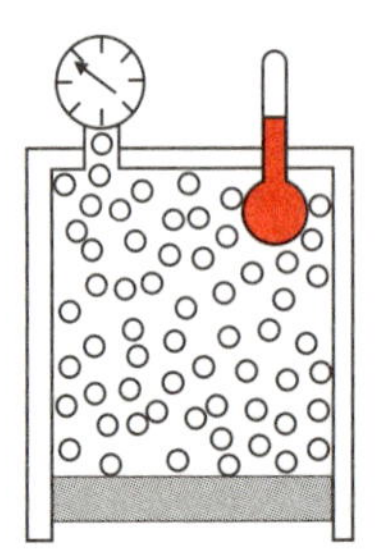
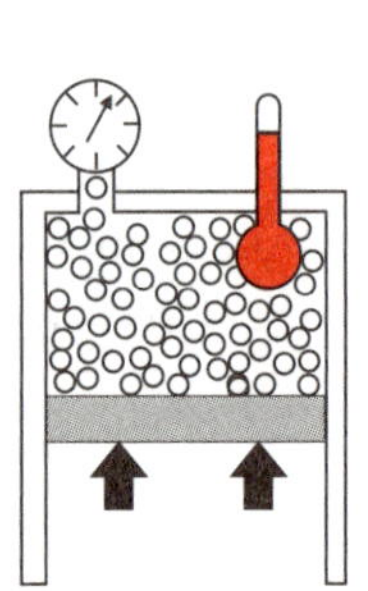
Figure 3-16 Compression of a gas raises its temperature.

And after a time the increased temperature from the compression process will dissipate through container walls. This results in air at a higher pressure than, but the same temperature as, the ambient of surrounding air.

In a pneumatic system, a positive displacement compressor delivers air to a system at an elevated pressure and temperature. But before the air is stored in an air receiver, it gives up this excess heat energy while flowing through the intercooler (in multistage compressors) and the aftercooler, service and distribution lines and at the same time has a pressure drop. This excess heat energy, which is a result of the compression process, will be removed by the system and probably will not be seen at the eventual point of work and is, therefore, an efficient part of a system (lost energy).

Friction

Another condition that adds to an inefficient system is friction. Friction is present between elements which are moving in relation to one another. In a pneumatic system, air is moving with respect to the pipe containing it. The faster air travels through a system, the more pressure energy is changed into heat.

The change of pressure energy into heat energy occurs in any dynamic or moving fluid power system as the frictional resistance of the system is overcome. However, we do not encounter pneumatic systems running hot. The reason for this is that air cools as it expands. Cooling from air expansion has more temperature effect on a pneumatic system than frictional resistance.

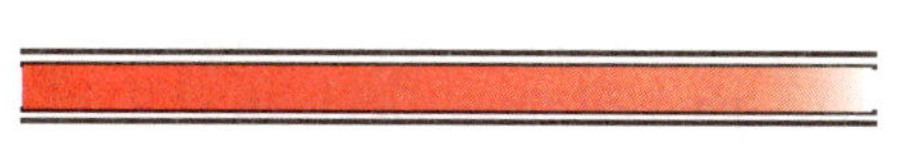

Figure 3-17

Changing Fluid Direction

One condition that adds greatly to system inefficiency is the need for the fluid to change its direction of flow. A mainstream of air molecules crashes into conductor surfaces and other air molecules as it is forced to change direction because of a pipe bend or elbow. This causes some pressure energy loss. Sometimes this is referred to as a friction loss. Many times this loss is expressed in terms of equivalent feet of straight pipe. An example would be a standard 2" ell. The rated loss is equivalent to 3.6 feet of straight pipe. Because any amount of pressure loss is so important in a typical industrial pneumatic system (pressure to 80-150 psi only) it should be obvious that the use of elbows (energy loss) be kept to a minimum.

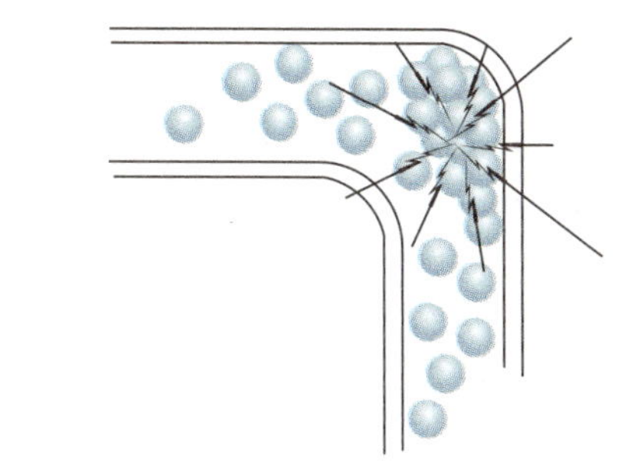

Figure 3-18 Fluid changing direction causes pressure energy loss.

Flow Rate

The volume of air flowing through a pipe in a period of time is a rate of flow. Flow rate in pneumatic systems is usually measured in cubic feet per minute (cfm). We know that a cubic foot of air can be under various pressure. A cubic foot of air can be at 90 psi, 100 psi, or 60 psi. But, the cubic foot of air referred to in the term cfm generally indicates a cubic foot of air at the vicinity of the compressor intake. After this air is compressed, it will have a smaller volume depending upon how much it has been compressed. As this reduced volume passes through system piping, it is still referred to as a cubic foot of air space since that is what it is under normal conditions.

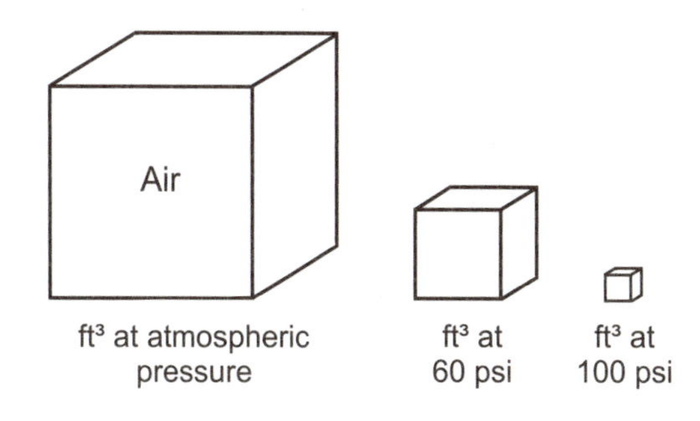

Figure 3-19

Free Air vs. Standard Air

Many times there is confusion about the terms "free air" and "standard air". Free air is a term which refers to the condition of the air supply available to a compressor. This is the air of the ambient or surrounding air at the compressor's inlet. Since atmospheric conditions vary from day to day and from place to place, free air varies widely in its properties and this is not a good term to use when comparing flow rates between various systems or for flow ratings of components. For this reason, cubic feet of free air is usually converted mathematically to cubic feet of standard air. A standard cubic foot of air is defined as air at a barometric pressure 29.92" Hg. (this is "sea level" pressure) with a temperature of 68°F and a relative humidity of 36%.

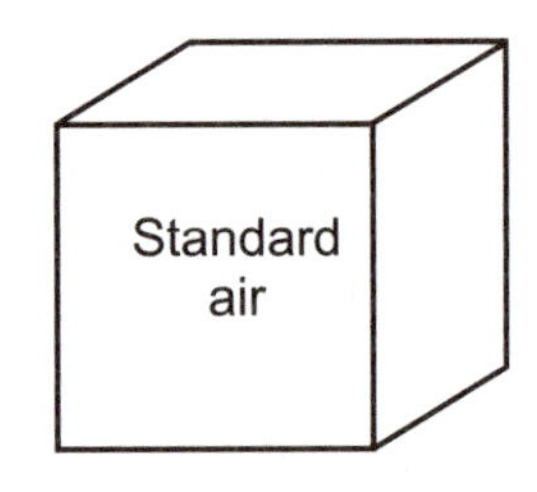

Figure 3-20

Velocity

In a moving dynamic system, air flowing through a pipe is traveling at a certain speed. This is the fluid's average velocity at any cross section and is usually measured in feet per second. The relationship between velocity and flow rate can be seen in Figure 3-21. In order to fill the tank with 20 cubic feet of standard air in one minute, the air must travel at a speed of 100 cubic feet per second. In a smaller pipe, 20 cubic feet of standard air must travel at a speed of 200 feet per second. In both cases the flow rate is 20 scfm. The fluid velocities, however, are quite different.

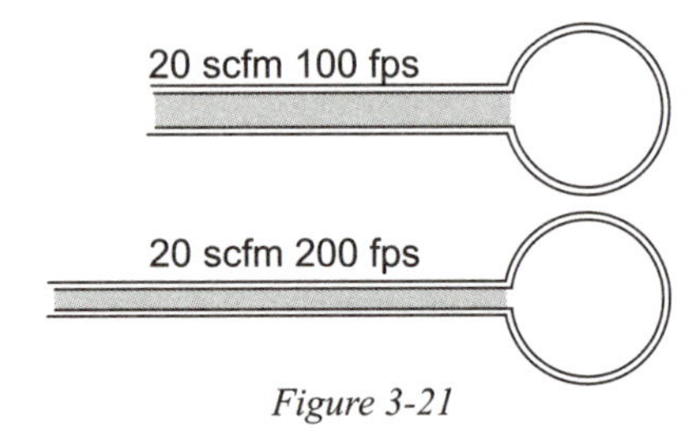

Figure 3-21

Critical Velocity

In a pneumatic system air is brought to an initially higher energy state by action of a compressor. The compressed air is then stored as potential energy in an air receiver tank. Flow discharges from the air receiver when there is a pressure difference between the initial state of the air in the receiver and the work requirement at the cylinder or motor.

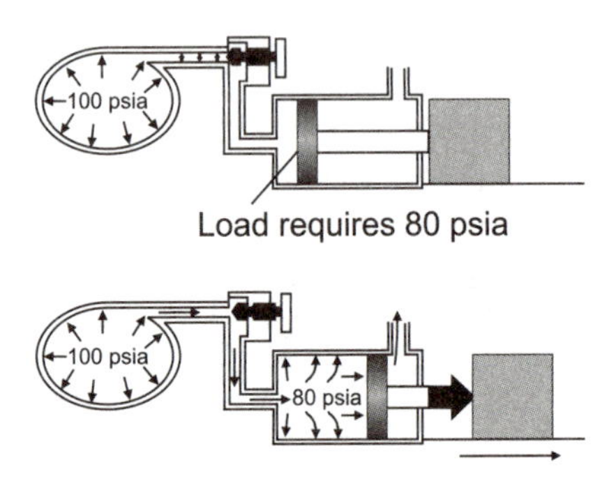

Figure 3-22

The maximum velocity at which air exists in an air system is called the critical velocity. This velocity is the speed of sound or about 100 fps at normal temperatures. This would be of great important when response times were critical. But such is not the case in the typical industrial pneumatic system. Therefore, any further discussion would be beyond the scope of this text and would be more academic than useful.

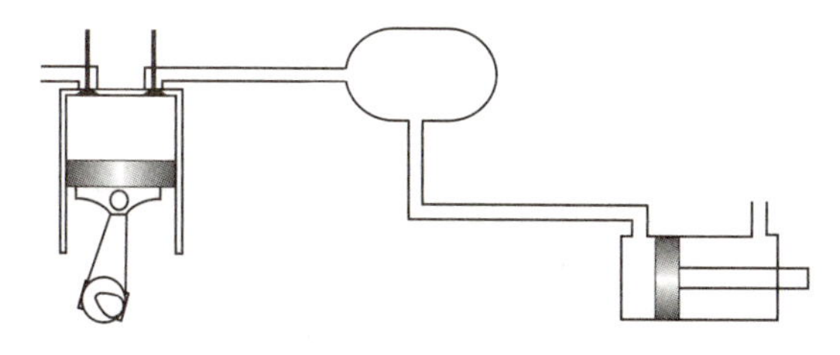

Figure 3-23

Pneumatic System Design

Industrial pneumatic systems are designed to operate with a minimum of frictional resistance. Working pressures for many pneumatic systems are around the 90 psi (6 bar) range. This energy must be used with care. An efficient industrial pneumatic system is designed with correctly sized pipes and components where bends and elbows are kept to a minimum so that pressure energy is not unnecessarily wasted.

An efficiently designed system also takes into account a compressor of appropriate size and type.

Lesson Review

In this lesson dealing with energy transmission using a pneumatic system, we have seen that:

- A gas is a substance made up of molecules which are continuously moving through space like a swarm of angry bees.
- Gases take the shape and volume of their container.
- Heat energy in a gas is molecular movement. A gas with increased temperature has faster moving molecules than a gas which is cool.
- Gas pressure is the result of millions of molecular collisions per second.
- When a trapped volume of gas has its volume reduced, its temperature and pressure increase.
- When a volume of a given gas expands, its temperature and pressure are reduced.
- In a pneumatic system, fluid flow occurs when air is allowed to expand.
- A positive displacement compressor delivers compressed air to an air receiver tank.
- Inefficiency in a pneumatic system is generally the result of heat during air compression, friction and fluid changing direction.
- Flow rate in a pneumatic system is generally measured in cubic feet per minute (cfm)/cubic decimeters per second (dm^3/s).
- Free air is a term which refers to the ambient air supply available to a compressor.
- Standard air is defined as air at sea level with a temperature of 68°F and a relative humidity of 36%.
- Fluid velocity in a pneumatic system is generally measured in feet per second (fps) (mls).
- Critical velocity of 100 fps (358 mls) in a pneumatic system is of great importance when response times are critical (53% differential).
- Pressure differential in a pneumatic system is simply the difference in pressure energy between any two points of the system.
- Pneumatic energy must be used with care.
- An efficient system design would include correctly sized pipes and components with bends and elbows kept to a minimum.

Exercise
Energy Transmission Using a Pneumatic System
50 Points

Instructions: For each incomplete sentence, one choice will correctly complete the statement. After reading the sentence and the possible choices, circle the letter next to the most **correct answer**. After all statements have been completed, place the letter for each answer in the appropriately labeled box below. The letter combination should form a word.

1. When a gas volume is reduced as in a compressor, gas _____ increases and temperature _____ .
   R. pressure, doubles
   M. potential energy, decreases
   E. pressure, increases
   A. potential energy, remains the same.

2. A hot gas has _____ molecules than a cool gas.
   A. more
   O. slower moving
   V. smaller
   B. faster moving

3. In a pneumatic system, the potential energy of compressed air is transformed into _____ .
   B. kinetic energy and fluid velocity
   R. kinetic energy and heat
   T. flow and kinetic energy
   Y. flow, air expansion

4. Flow rate in a pneumatic system is expressed as _____ .
   S. cfm
   G. gpm
   P. psia
   F. fps

5. When a gas volume is expanded, gas pressure _____ and _____ decreases.
   M. doubles, potential energy
   H. remains the same, potential energy
   D. increases, temperature
   P. decreases, temperature

6. In a pneumatic system, fluid flow occurs when air is allowed to _____ .
   M. expand
   A. cool
   E. heat
   K. compress

7. Inefficiency in a pneumatic system is generally the result of _____ .
   L. friction and heat of compression
   N. friction
   S. fluid changing direction and friction
   E. friction, fluid changing direction and heat of compression

8. _____ air is defined as air at sea level with a temperature of 68°F and a relative humidity of 36%.
   O. Ambient
   T. Standard
   E. Free
   C. Normal

9. Critical velocity in a pneumatic system is achieved when downstream pressure reaches _____ of upstream pressure.
   U. 47 psi
   R. 53 psi
   E. 53%
   J. 47%

Answer 4	Answer 9	Answer 5	Answer 8	Answer 1	Answer 6	Answer 2	Answer 7	Answer 3

Chapter 4
Control of Pneumatic Energy

Working energy that is transmitted through a pneumatic system must be directed and under complete control at all times. If not under control, useful work may not be done and machinery or machine operators might be harmed. One of the advantages of transmitting energy pneumatically is that energy can be controlled relatively easily.

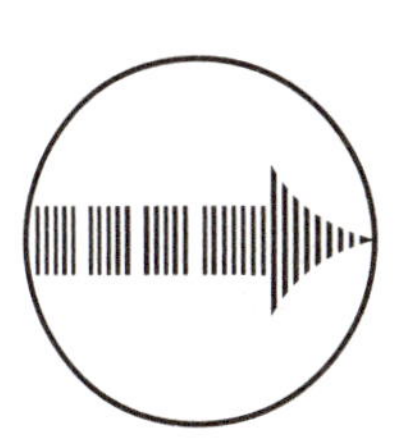

Figure 4-1

Valves

The basic pneumatic valve is a mechanical device consisting of a body and a moving part which connects and disconnects passages within the body. The flow passages in pneumatic valves carry air. The action of the moving part may control system pressure, direction of flow and rate of flow. Let us first look at pressure control.

Control of Pressure

Pressure in a pneumatic system must be controlled at two points – at the compressor and after the air receiver tank. Control of pressure is required at the compressor as a safety for the system. Control pressure at the point of air usage is necessary for safety and so that actuators receive the proper pressure to avoid wasting energy.

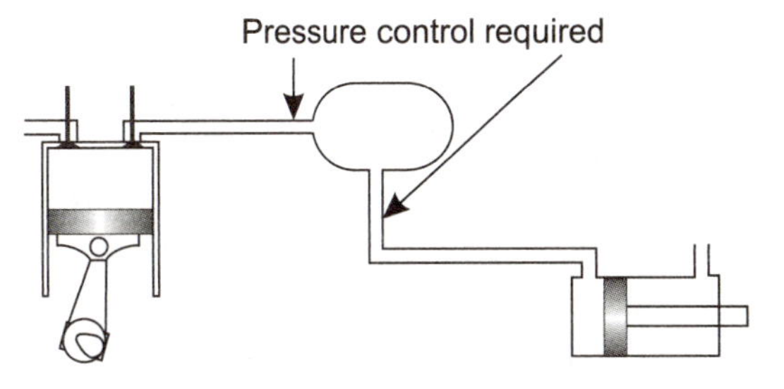

Figure 4-2

Control of Pressure After a Compressor

Characteristically in a pneumatic system, energy delivered by a compressor is not used immediately, but is stored as potential energy in an air receiver tank in the form of compressed air. In most applications, a compressor is designed into a system so that it operates intermittently. A compressor usually delivers compressed air to a receiver tank until high pressure is reached, then its air delivery is controlled. When air pressure in the tank decreases, the compressor delivery is increased (within its design limits). Compressor operation in this manner is a power savings for the system. One way to accomplish this function is through the use of a pressure switch.

Pressure Switch

A common way of sensing tank pressure and controlling actuation and de-actuation of a relatively small (5-10 hp) compressor is with a pressure switch. Large compressors use other means of various types.

How a Pressure Switch Works

In a pressure switch the system pressure is sensed at the bottom of the piston through the pressure switch inlet. When pressure in the system is at its low level, the spring pushes the piston down. In this position, a contact is made causing an electrical signal to turn on the compressor.

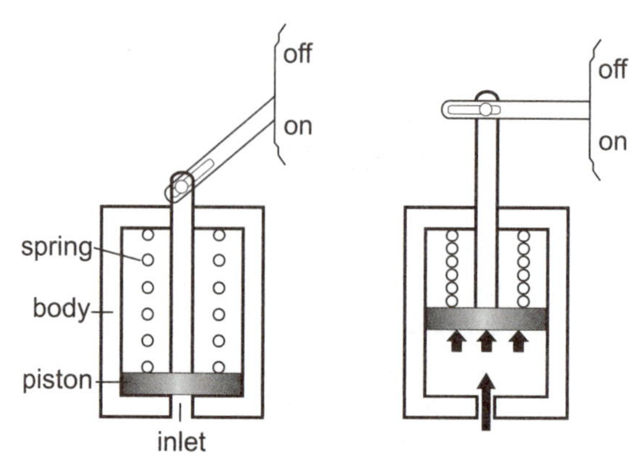

Figure 4-3

As pressure in the receiver tank rises, it forces the piston upward. With system pressure at its high level, the piston breaks the electrical contact, shutting down the compressor. However, it should be noted that in large compressors, the compressor is not shut off. In this type the compressor is unloaded in some way to prevent the compression stroke from compressing any air.

Safety Relief Valve

However, maximum pressure developed by a compressor is designed to be regulated by a control system which senses discharge or tank pressure. In case of an emergency, as in the failure of a control system to function properly, a positive displacement compressor system should have a safety relief valve. This is mandated by law for most applications.

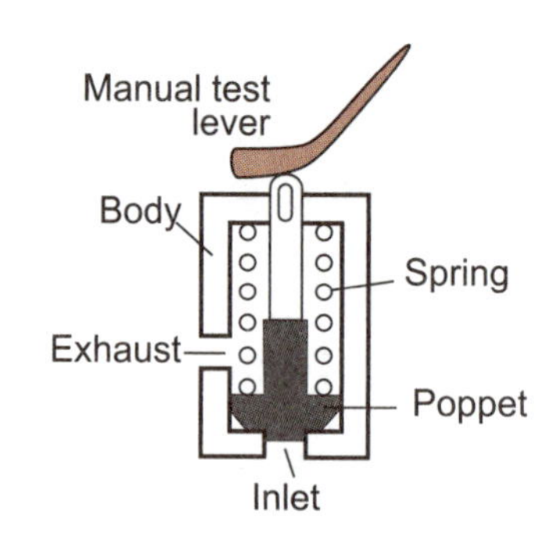

Figure 4-4 Safety relief valve with manual test lever

A safety relief valve consists of a valve body with inlet and exhaust passages and an internal moving part. In figure 4-4, the moving part is a poppet which is held biased in one position by a spring. To comply with safety codes the valve will have a manual test lever.

How a Safety Relief Valve Works

The safety relief valve is a normally non-passing (closed) valve. The poppet of the safety relief valve is seated on the valve inlet. A spring holds the poppet firmly on its seat. Air cannot pass through the valve until the force of the spring biasing the poppet is overcome. Air pressure at compressor outlet is sensed directly on the bottom of the poppet. When air pressure is at an undesirably high level, the force of the air on the poppet is greater than the spring force. When this happens the spring will be compressed and the poppet will move off its seat and air will exhaust through the valve vent (exhaust) ports.

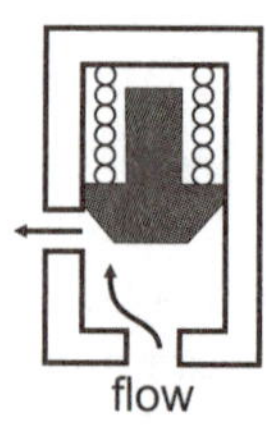

Figure 4-5

Safety relief valves on a compressor are not intended to be operated frequently. They are a safety device. Many times safety relief valves are equipped with whistles or horns to alert personnel that something has failed or a problem exists. Good maintenance and safety procedures require the periodic checking of these safety valves to verify that the valve can move freely. Many times this can be done by moving the manual test lever to open the valve.

Pressure Regulator

Next we must control pressure downstream of the receiver tank. In a pneumatic system, energy which will be used by the system and transmitted through the system is stored as potential energy in an air receiver tank in the form of compressed air. A pressure regulator is positioned after a receiver tank and is used to control the pressure of this stored energy to a leg of a circuit.

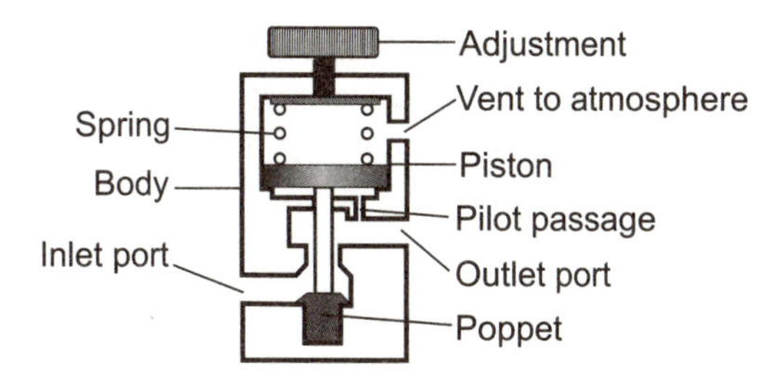

Figure 4-6

A pressure regulator valve consists of a valve body with inlet ("primary") and outlet ("secondary") passages. The regulator has a moving part which controls the size of the opening between the primary and secondary passages. In our illustration, the moveable part is a poppet which is connected to a piston. This piston is biased away from its seat ("open") by an adjustable spring.

How a Pressure Regulator Works

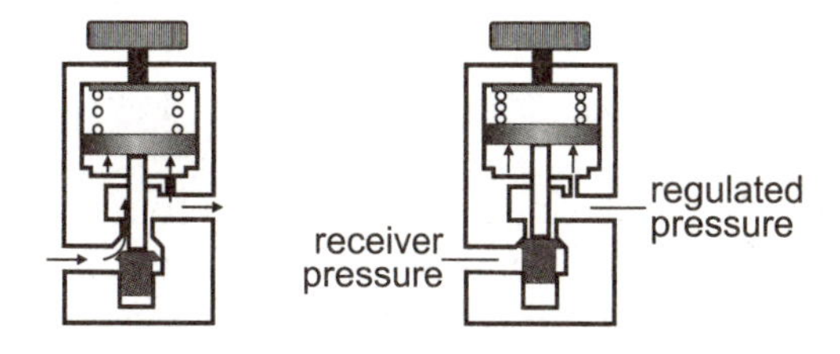

Figure 4-7

The pressure regulator is a normally passing (open) valve. "Open" means that the flow passage normally allows air to flow freely. With a regulator positioned after a receiver tank, air from the receiver can expand (flow) through the valve to a point downstream of the secondary passage. When pressure in this passage of the regulator rises, it is transmitted through an internal pilot passage which leads to the piston area on the opposite side of the spring. The piston has a relatively large surface area exposed to secondary pressure and is therefore responsive to secondary pressure fluctuations. When the controlled pressure nears the preset level, the piston moves upward allowing the poppet to move towards its seat to control (or throttle) the flow. The poppet blocks flow once it seats and does not allow pressure to continue building downstream. In this way, air at a controlled pressure is made available to an actuator downstream. Also, a pressure regulator insures that energy in an air system is not wasted.

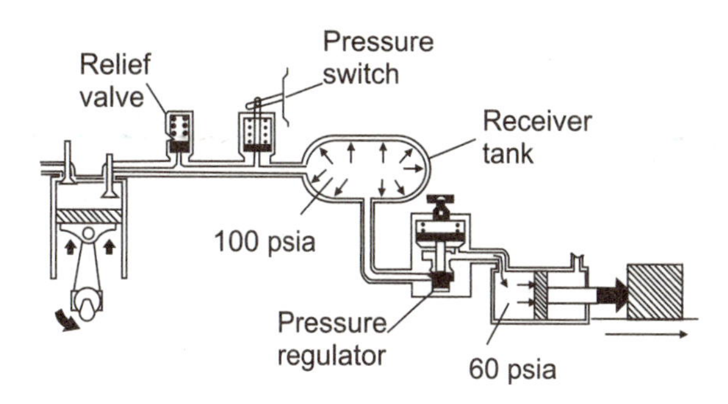

Figure 4-8

For instance, if an air receiver charged to 100 psia were allowed to subject its full pressure to an actuator which only requires 60 psia, pressure energy would be wasted. A regulator that is set to deliver only the amount of pressure needed to overcome load resistance plus a few psi additional to develop flow should be positioned before the actuator.

Control of Actuator Direction

Once a cylinder is extended, it has to be retracted. To perform this task, a valve is used to change the direction of air flow to and from the cylinder. This is generally done by using a double-acting cylinder and a directional control valve.

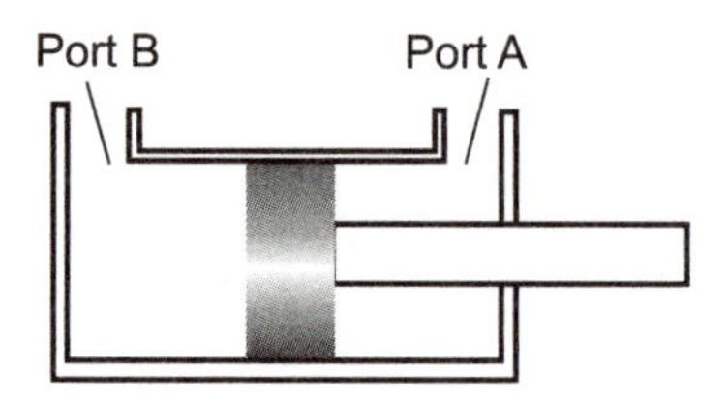

Figure 4-9

Double-Acting Cylinder

The double-acting cylinder has a port at each end of the cylinder body by which air under pressure can enter or exhaust. This causes the piston rod to extend or retract (double-acting). To distinguish the ports on a double-acting cylinder, we will label one A and the other B.

Directional Control Valve

Next, to change the direction of air flow to and from the cylinder, we use a directional control valve. The moving part in a directional control valve will connect and disconnect internal flow passages within the valve body. This action results in a control of air flow direction.

The typical directional control valve consists of a valve body with four internal flow passages and a moving part, a spool, which alternately connects a cylinder port to the supply pressure or to exhaust.

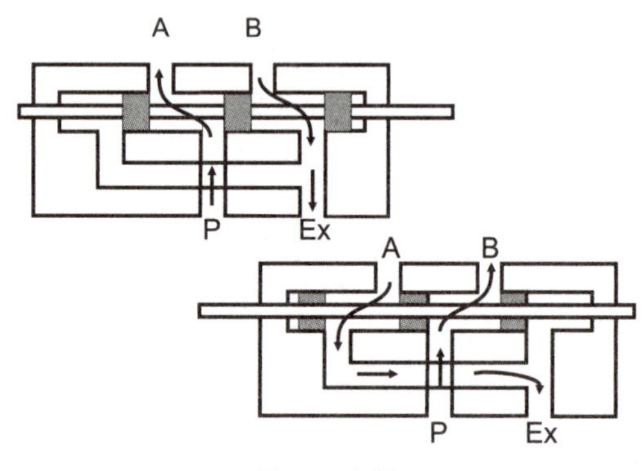

Figure 4-10

How a Directional Control Valve Works

With the spool in one extreme position, supply pressure is connected to port B and port A is connected to exhaust. If cylinder port A is connected to the valve port A and cylinder port B to the valve port B, then with the spool in the other extreme position, supply pressure is connected to cylinder passage A and cylinder passage B is connected to exhaust. With a directional control valve in a circuit, the cylinder's piston rod can be extended or retracted and work performed. If we control how fast the work is performed, we must control the flow rate.

Control of Flow Rate

In a pneumatic system, actuator speed is determined by how quickly the actuator can be filled and exhausted of air. In other words, speed of a pneumatic actuator depends upon the force available from the pressures acting on both sides of the piston, a result of cfm flowing into the inlet and out of the exhaust port.

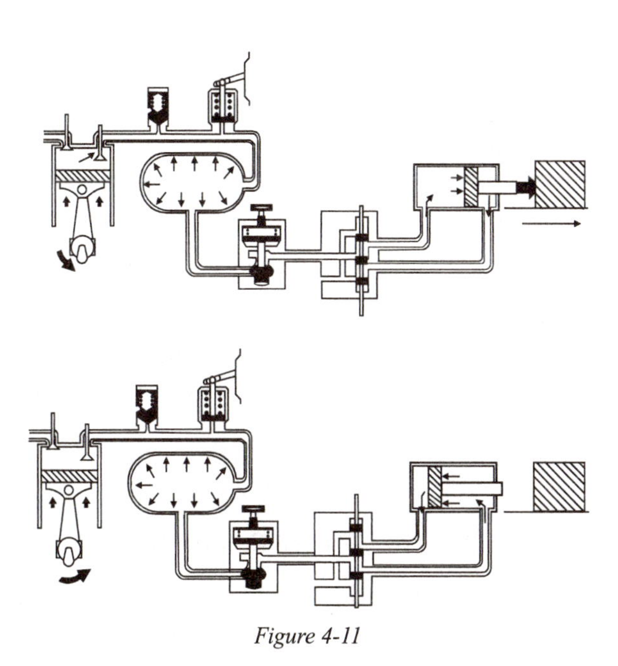

Figure 4-11

We have seen that a regulator will influence actuator speed by portioning out to its legs of a circuit the pressure required to equal the load resistance at an actuator. This additional pressure is used to develop air flow. Even though this is the case, pressure regulators are not used to vary actuator speed. In a pneumatic system, actuator speed is affected by a restriction, as with a needle valve or a needle valve with a bypass check, often times called a "flow control" valve. This valve will meter flow at a constant rate only if the system resistance, the total load (friction, load) are constant and the pressure at the inlet of the cylinder and at its outlet do not vary through out the entire work cycle. This valve does not control flow, it only affects flow. An example of this would be when the load encounters an added resistance, its speed would decrease or stop.

Safety note: Due to the compressibility of air, speed control circuits must be examined very closely for all possible failure conditions.

Once the resistance was overcome, there would be a very rapid increase in speed. This could be a very dangerous condition for the operator or nearby personnel.

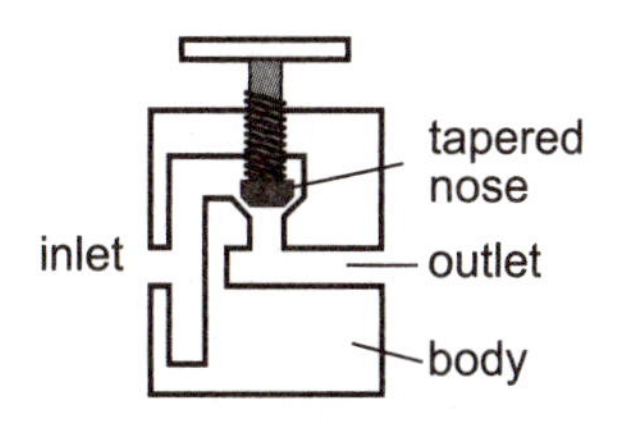

Figure 4-12 Needle valve components

Needle Valves

As indicated earlier, a needle valve in a pneumatic system affects the operation by causing a restriction.

The typical needle valve consists of a valve body and an adjustable part. The adjustable part in our example is a tapered-nose rod which is threaded into the valve body.

The part in this valve is described as adjustable because it can be repositioned or "adjusted" fairly easily.

How a Needle Valve Works

In order to reduce the speed of an actuator in a pneumatic system, a needle valve is sometimes positioned in a circuit so that it restricts air flow exhausting from an actuator port. The more the tapered-nose rod is screwed towards its seat in the valve body, the greater the restriction to free flow.

By restricting exhaust air flow in this manner, a back-pressure is generated within the actuator, thus reducing the forces available to create motion. This means a larger portion of regulator pressure is used to overcome the resistances at the actuator and less pressure energy is available to develop flow.

With less cfm flowing into the actuator, actuator speed decreases.

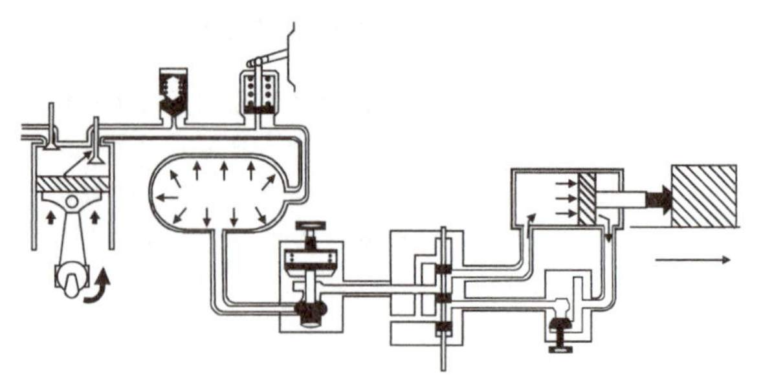

Figure 4-13

By controlling the amount of restriction developed by a needle valve, the speed of an actuator can be "controlled", **but only if the total load is constant.**

Now let us build a simple pneumatic system.

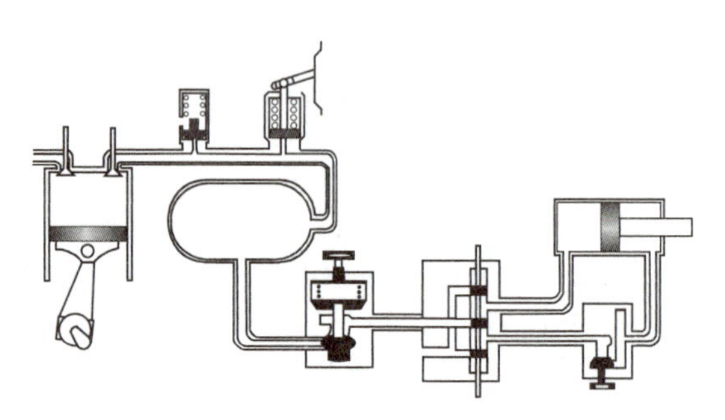

Figure 4-14 A simple pneumatic system

A Simple Pneumatic System

When all of the pneumatic components which have been described to this point are put together in the proper order, they make up a simple pneumatic system. The system can perform useful work because pneumatic working energy in the system can be controlled. All pneumatic systems operated on the same principles which have been discussed up to now. In the remaining text material dealing with pneumatics, we shall concentrate on some of the typical components which are readily available for industrial pneumatic systems. Before we proceed, a word about pneumatic symbols is needed.

Pneumatic Symbols

The pneumatic components and elementary systems which have been shown to this point have been illustrated in a pictorial manner. System diagrams have been shown with simplified component cutaways to illustrate integral component operation. This technique is beneficial from an instructional viewpoint, but it is impossible from a workaday standpoint. In pneumatics, just as in other technologies, symbols are used to describe components and systems. The symbols for the component which have been discussed and the simple system which has been developed are illustrated using graphic symbols for fluid power.

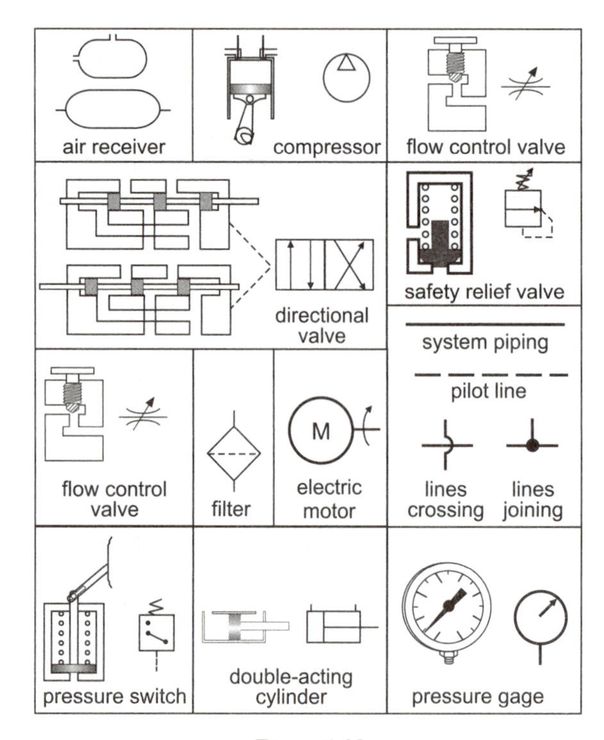

Figure 4-15

Note: In addition to the components which have been discussed, the system consists of many others, such as air line lubricators, mufflers, oil reclassifiers, etc. Pneumatic systems are generally equipped with a prime mover like an electric motor. And, in order to achieve a degree of reliability, pneumatic systems should be protected from dirt with an air filter.

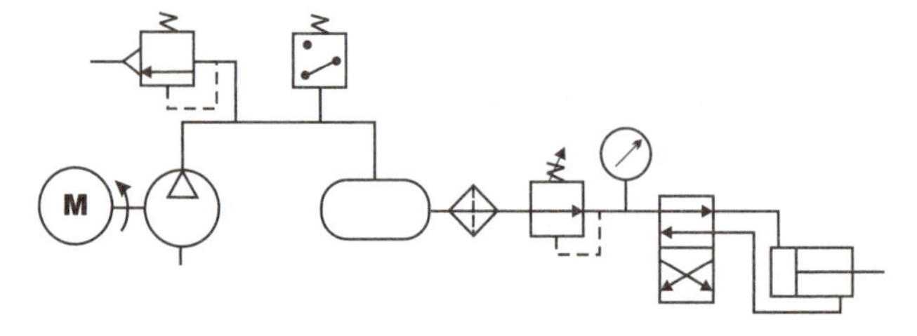

Figure 4-16 Pneumatic schematic

Lesson Review

In this lesson dealing with the control of pneumatic energy, we have seen that:

- Pneumatic energy is controlled with valves.
- A valve is a mechanical device consisting of a valve body and an internal moving part which connects and disconnectes passages within the body.
- Pressure in a pneumatic system must be controlled at two points – after the compressor and after the air receiver.
- Compressors in pneumatic fluid power systems are usually designed to operate intermittently.
- A pressure switch is a common way of sensing pressure and controlling actuation and de-actuation of a small compressor.
- In case of an emergency, a positive displacement compressor is equipped with a relief valve.
- A pressure regulator delivers a constant pressure air supply to its leg of a circuit and insures that the stored pressure energy of an air receiver is not unnecessarily wasted.
- A directional valve controls the direction of air flow to and from an actuator.
- A double-acting cylinder is often used in a pneumatic system to achieve a reciprocating motion.
- In a pneumatic system, actuator speed is determined by how quickly the actuator can be filled and exhausted of air.
- A flow control valve operates by causing a restriction.
- Symbols are used to describe components in pneumatic systems.

Exercise
Control of Pneumatic Energy
50 Points

Instructions: In this assignment the answers are already given. You select a question at the end of the exercise which most correctly satisfies an answer.

Answers:

_______ 1. A normally closed condition

_______ 2. After a compressor and after an air receiver

_______ 3. The speed at which an actuator is filled and exhausted of air

_______ 4. It operates by causing a restriction

_______ 5. Direction of flow, rate of flow and maximum pressure

_______ 6. Cuts in and cuts out a compressor

_______ 7. Under complete control

_______ 8. A pressure regulator

_______ 9. A body and an internal moving part

_______ 10. A normally open condition

Questions:

A. In what condition is a pressure regulator found?

B. What function does a pressure switch perform in a pneumatic system?

C. What insures that the stored pressure energy of an air receiver is not unnecessarily wasted at an actuator?

D. Where must pressure be controlled in a pneumatic system?

E. What determines an actuator's speed of doing work?

F. How does a flow control valve in a pneumatic system work?

G. How must energy in a pneumatic system always be used?

H. What do valves control in a pneumatic system?

I. What do all valves basically consist of?

J. In what condition is a safety relief valve found?

Chapter 5
Compressors

Compressors convert the mechanical energy provided by a prime mover (e.g. electric motor, internal combustion engine, etc.) into the potential energy of compressed air.

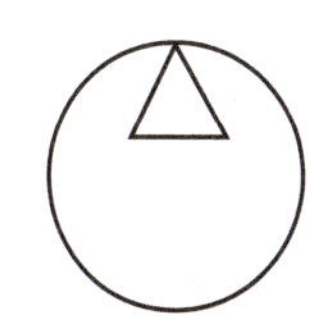

Figure 5-1 Compressor symbol

To accomplish useful work with a pneumatic system, a device is needed which can supply a sufficient amount of air at a desired pressure. This device is a compressor.

Compressor Types

There are two basic groups of air compressors:
1. displacement
2. dynamic

In the displacement type compressors, pressure increases because of the change in the volume of air trapped in a confined space. A positive displacement piston type compressor fits into this category. In this type of unit, capacity is unaffected by changes in working pressure (neglecting leakage and volumetric efficiency).

However, in the dynamic type compressor, pressure rise is caused by adding kinetic energy to accelerate the moving gas and converting the velocity energy to pressure energy in some sort of diffuser. Let us briefly look at several types of air compressors in more detail.

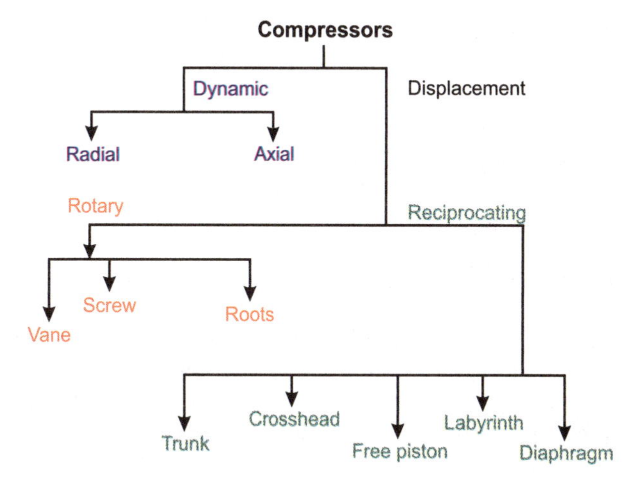

Figure 5-2

Positive Displacement Compressors
Piston Type Compressors

The most common type of positive displacement compressor found in industrial pneumatic systems is a reciprocating piston compressor.

A simplified example of a reciprocating piston compressor is illustrated. This type of compressor is basically a piston inside of a bore. The piston is connected to a crankshaft which in turn is connected to a prime mover. Two valves control inlet and outlet flow through the compressor.

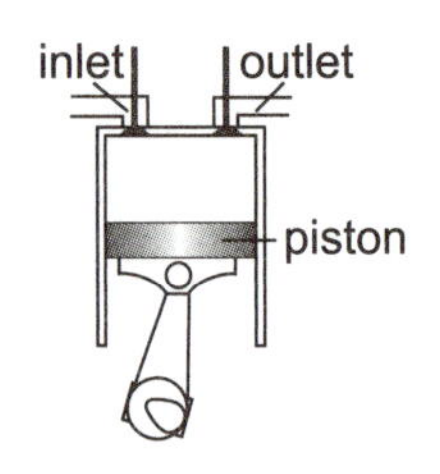

Figure 5-3 Single-stage compressor

How a Reciprocating Piston Compressor Works

As the crankshaft of a reciprocating piston compressor rotates, the piston moves within the bore. When the crankshaft pulls the piston in one direction, an increasing volume is formed within the bore. With the resulting less-than-atmospheric pressure and the intake valve open, atmospheric air fills the chamber. At the end of the piston stroke, the chamber is filled with air and the intake valve closes.

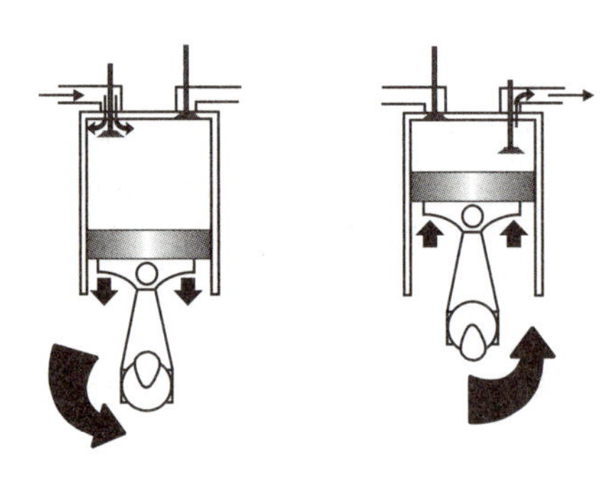

Figure 5-4

Starting its upward travel, the piston compresses the air. When air pressure in the bore reaches a high level, the outlet valve is opened. Compressed air discharges from the compressor into an air receiver tank.

This describes the operation of a single-stage piston compressor, that is, a compressor which compresses air in a single stroke before it is discharged. Single-stage compressors (with a much larger compression ratio capability than the example shown) are generally used in systems which require compressed air between 40 and 100 psig. However, there are some single-stage compressors capable of 250 psig.

We learned earlier that when a gas is compressed, as in a compressor, the gas heats up. In a single-stage compressor, when air is compressed above 80 psi, the heat created (heat due to air being compressed) and the energy required to compress the hot air becomes excessive. However, many industrial systems operate above 80 psi. For this reason, two-stage compressors are usually found in industrial pneumatic systems. This type of compressor will be discussed later.

There are many different designs used in piston type compressors. They may be either lubricated or nonlubricated. Nonlubricated models have piston rings or skirts composed of low friction materials, e.g. graphite or Teflon™.

The delivery from a reciprocating type compressor may have severe pulsations especially if it is a single cylinder device closely coupled to its load. To lessen these pulsations, a receiver tank is installed to act as a smoothing accumulator. This may be necessary even if the output is directly matched to the load cfm requirement of the system. There are other types of positive displacement compressors. Another is the vane type compressor.

Vane Compressors

Vane compressors generate a pumping action by causing vanes to track along a circular housing.

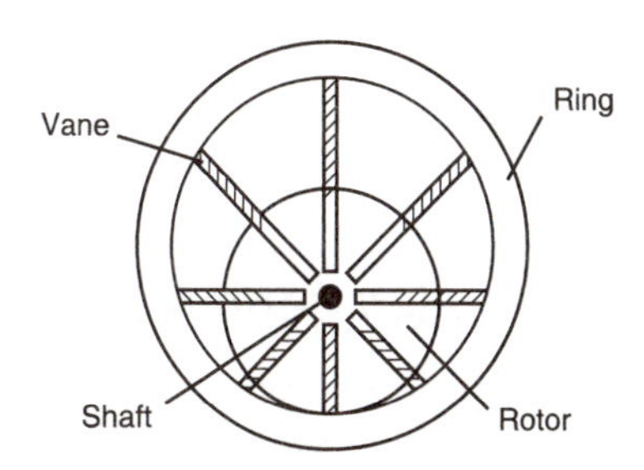

Figure 5-5 Schematic with exaggerated eccentricity for emphasis

What a Vane Compressor Consists Of

The compressing mechanism of a vane compressor basically consists of housing, rotor and a series of floating vanes. The vanes are typically made of carbon, cloth impregnated with phenolic resin or a similar rigid low friction material For oil free models, bronze and carbon vanes are often used.

How a Vane Compressor Works

The rotor of the vane compressor houses the vanes and the rotor is attached to a shaft which is connected to a driver motor. As the motor turns the rotor, the vanes are thrown out by centrifugal force and track along the housing. As the vanes make contact with the housing, a seal is formed between vane tip and ring.

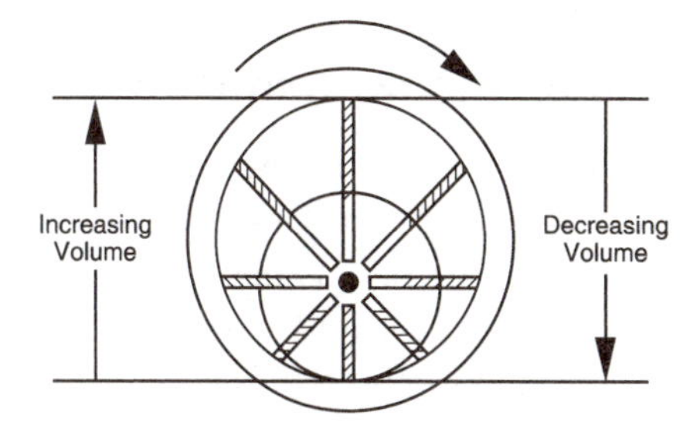

Note: Ring does not rotate.

Figure 5-6

The rotor is positioned off center to the housing center. As the rotor is turned, an increasing and decreasing volume is formed within the housing.

This type of compressor is relatively low in price for small displacement, provides low operating cost and requires a low starting torque. They commonly range in pressure up to 150 psig (10 bar) with power ranges from fractional hp to 500 hp (373kW). A more sophisticated type of compressor is the helical compressor.

Helical Compressors

A helical compressor generates compressed air by running two meshing rotors on one another like two screws. This compressor is many times called a "screw" type compressor.

Figure 5-7 Helical compressor

What a Helical Compressor Consists Of

The compressing mechanism consists of two helical screws working much like a hydraulic screw pump. Air is drawn in from one end and exits the other. There are two basic types: oil flooded or dry. Typically, the maximum pressure for single-stage units is 125 psig (9 bar).

Dry Helical Compressors

The first type, dry helical, uses timing gears to turn the two screws. This maintains a constant clearance which means lubrication is usually not needed. These units have a high efficiency when run at high rpm. Today, their use is rare except where oil-free air is needed. However, improvements have boosted the screw compressors' efficiency greatly, as much as 15%. These improvements have spotlighted the oil-flooded compressor.

Oil Flooded Compressors

In this type of compressor, timing gears are not needed but lubrication is because the helixes are run on each other. Efficiency is higher because the lubricating oil provides a good air seal between the two rotating members and the case. However, oil separators are needed to remove the oil from the air downstream of the compressor. The range of power for these units is 7 to 300 hp (5.2 to 223 kW).

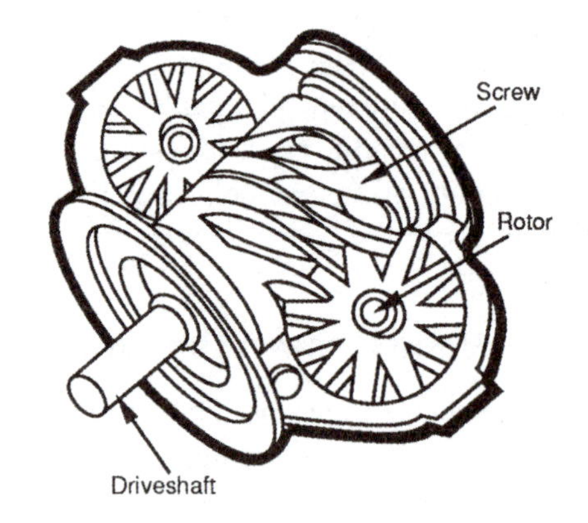

Figure 5-8 Single-screw compressor (Courtesy Penton/IPC Publication - Machine Design)

Single Screw Compressors

There is a modification of the screw compressor called the single screw compressor. This design is based upon the helical compressor but, as the name implies, only one screw exists. As the screw in the center rotates, air trapped between the screw teeth is compressed against the rotors. Air compressors of this type have low vibration and noise levels and typically, pressures are less than 50 psi (3.5 bar).

Lobed-Rotor Compressor

Still another type of compressor is the lobed-rotor compressor. This constant displacement compressor is very much like the dry helical type. A timing mechanism is employed to eliminate problems with lobe and housing clearances. Because the rotors do not touch, a certain amount of "slip" exists. This "slip" increases as output pressure increases. This device should be operated at maximum speed for the highest efficiency. Pressures of up to 250 psi (17 bar) are obtainable with horsepowers between 7 and 3000 (5 to 2237 kW). Now let's look at the most common dynamic compressor, the centrifugal type compressor.

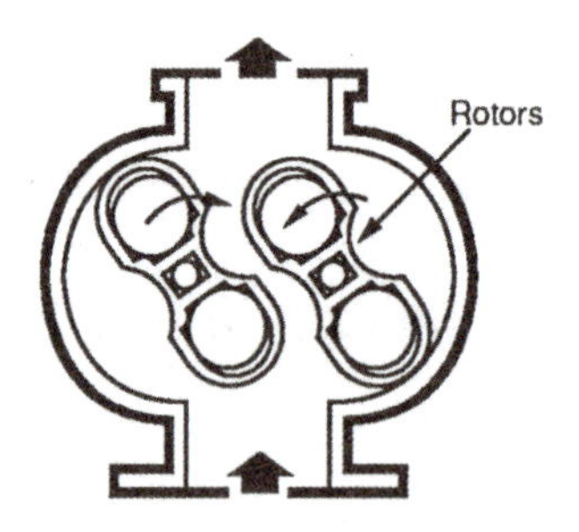

Figure 5-9 Lobed-rotor compressor (Courtesy Penton/IPC Publication - Machine Design)

Centrifugal Compressors

This is not a constant displacement type compressor. Centrifugal types are best suited for moving large volumes of air at relatively low pressures.

How a Centrifugal Compressor Works

This compressor delivers compressed air through the use of centrifugal force. The compressing mechanism consists of an impeller, a diffuser section where velocity energy is converted to pressure, followed by a collector where air velocity is further reduced again increasing pressure.

With centrifugal compressors, air demand should never be allowed to drop much below the rated flow. If this is allowed to happen, the compressor will be unloaded and will surge. In this condition, the pressure field at the compressor's outlet becomes unstable. If continuous operation in this condition occurs, bearings, blades and even the housing may be damaged.

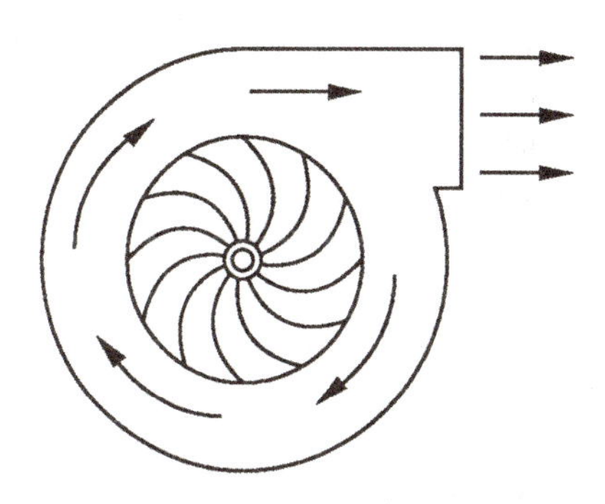

Figure 5-10 Centrifugal compressor

Radial Compressors (Centrifugal)

Where very high speeds are involved, the radial compressor should be used. Speeds in the area of 50,000 to 100,000 rpm are common in aircraft and aerospace industries where weight is also a problem. Most commercial units operate about 20,000 rpm. Typically, the minimum capacity of a centrifugal compressor is limited to the flow through the last stage. For example, the practical limit of a horizontal split type is 350 cfm (165 dm^3/s).

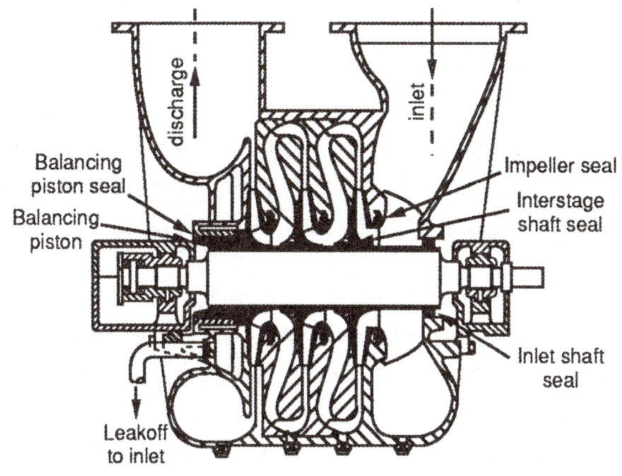

Figure 5-11 Vertical section drawing showing typical multistage centrifugal compressor. (Courtesy Compressed Air and Gas Institute)

Axial Compressors (Centrifugal)

Another centrifugal type compressor is the axial compressor. This type of compressor is characterized by having its flow move axially through the rotating group. This type has a smaller diameter than radial and revolves at higher speeds for the same output rate (about 25% faster). These are typically used for constant flow at moderate pressures (90 psi [6 bar] except for aircraft engines).

Axial compressors are best suited for high volumes of air. An example would be blast furnace blowing. Minimum capacities are usually above 135,000 cfm (63720 dm^3/s). Now that we have examined the most common types of compressors, let us discuss multi-stage compressors.

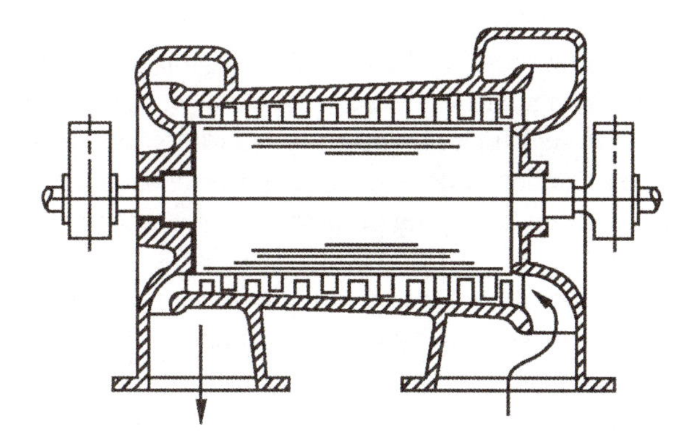

Figure 5-12 Multistage, single-flow axial compressor (Courtesy Compressed Air and Gas Institute)

Multi-stage Compressors

Because it is more efficient to compress a gas to higher pressures in numerous stages, many multi-stage compressors are offered. We will use a two-stage piston compressor for an example.

Two-stage Piston Compressors

A two-stage piston compressor basically consists of a large and small piston, each in their own cylinder bore, connected to the same crankshaft, with associated inlet and outlet valves and an intercooler.

How a Two-stage Piston Compressor Works

In a two-stage compressor, the piston with the large diameter performs the first stage of compression. The smaller piston compresses air in the second stage. As the crankshaft is turned by its prime mover, the large diameter piston strokes downward. Air enters the chamber from the atmosphere through the open inlet valve. When the piston starts its upward movement, the inlet valve closes. Air is compressed (and heated) until a certain pressure is reached, at which time the outlet valve opens, discharging hot, compressed air into the intercooler. If this air were discharged directly into an air receiver, the air would quickly cool to the temperature of the surrounding atmosphere losing pressure.

The compressed air is directed through the intercooler to the second stage piston. During the travel time from first to second stage, the air is cooled by means of air blowing over or water flowing across the intercooler. By the time the air reaches the second stage piston, a great portion of the heat of first stage compression has been dissipated. The air is now cooler and ready to be compressed a second time.

With compressed air at its inlet, the smaller diameter piston is pulled downward. Compressed air fills the chamber and the inlet valve closes. The piston is stroked upward, compressing the air again. The compressed air, as it discharges from the compressor, is at an elevated temperature. But this excess temperature above the ambient is not nearly as great as if a single-stage were used in the compressing process.

Two-stage compressors do not require as much energy in compressing air as single-stage units.

Even though these devices are quite efficient, their output must be controlled. If this is not done, an unsafe condition may exist and no productive work will be accomplished.

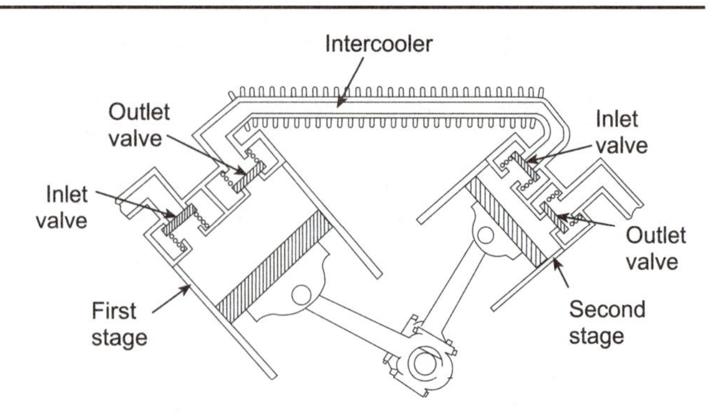

Figure 5-13 Two-stage compressor components

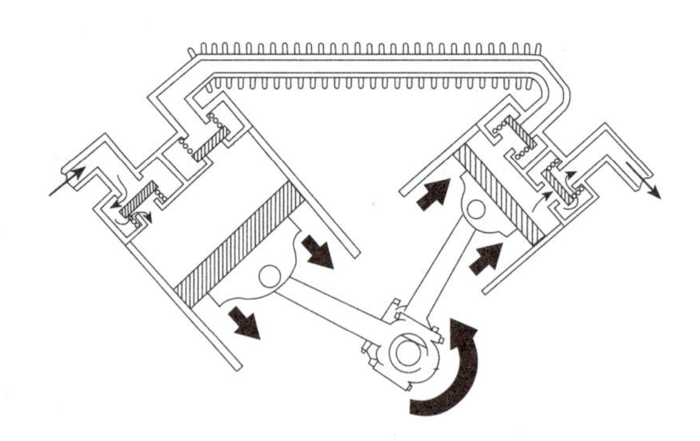

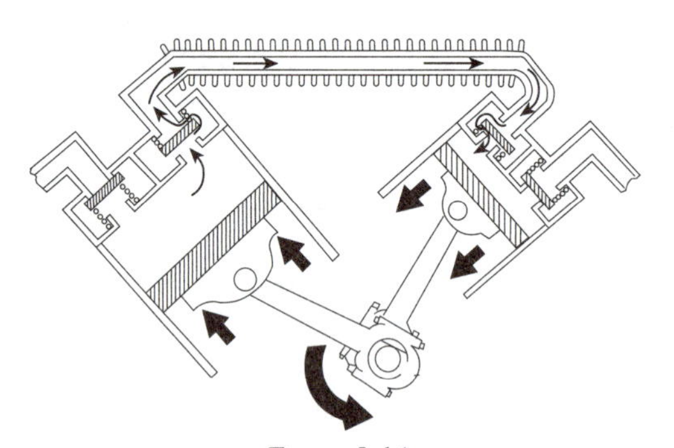

Figure 5-14

Capacity Control

The output capacity of a compressor must be regulated to match the demand of the system. The typical sensor used to control compressor output is a pressure control. It is typically automatic, but a manual means of unloading may also be available. Compressor unloading will also be necessary in situations where driver torque is not sufficient to accelerate the compressor under load.

Unloading Methods or Output Control

There are many ways of unloading or controlling the output of a compressor.

One method termed bypass will bleed the compressed air to the atmosphere at a high pressure. This type of control will cause the compressor to run under full load at all times. Under these conditions, efficiency is poor and care should be taken to provide cooling if the air is allowed to recirculate.

Another method is to start and stop the compressor as directed by the maximum and minimum pressure settings. This can be accomplished by starting and stopping the motor. However, frequent starting and stopping of an electric motor may cause early motor failure due to excessive heat buildup. Other means of accomplishing this would be with a clutch between the motor and compressor or with a variable speed drive. Still another method of unloading is inlet valve regulation. In this method a mechanism holds open the inlet valve when there is no demand for air. This allows the air to sweep in and out of the inlet valve as the piston moves in the bore. Another method, inlet throttling, increases or decreases the restriction at the inlet port of the compressor. By doing this, it also regulates the amount of air admitted to the cylinder, thus decreasing its capacity. The drawback of such a control is that the compression ratio increases thus maintaining a higher horsepower demand on the compressor motor than other methods of compressor unloading.

Total closure type control is seldom used today. With this method the inlet valve is continuously closed during unloading. This causes the cylinders to essentially run in a vacuum. The unloaded power consumption is proportioned to mechanical friction. All of these controls for unloading the compressor affect its performance and efficiency. However, they are not the only thing to do so. Another is high altitude.

Unloading Methods

1. Bypass
2. Start stop
3. Inlet valve regulation
4. Inlet throttling
5. Inlet closure

Effect of Altitude on Performance

The higher the elevation, the lower the air pressure density. This changes the power needed to drive the compressor, the compression ratio and compressor capacity. Probably of greatest concern to an engineer is the capacity reduction of the compressor due to altitude. A chart is given showing the reduction with altitude. To find the capacity of a single-stage compressor, multiply the capacity correction factor by the sea level capacity. It should be noted that due to the relatively low first stage ratio of compression in multi-stage systems, the free air capacity is affected only slightly as compared to single-stage units.

Capacity Correction Factors
(Estimate Gage Pressure - psi)

Altitude	40	60	80	90	100
0	1.0	1.0	1.0	1.0	1.0
1000	0.993	0.992	0.992	0.988	0.987
2000	0.987	0.984	0.977	0.972	0.972
3000	0.981	0.974	0.967	0.959	0.957
4000	0.974	0.963	0.953	0.944	0.942
5000	0.967	0.953	0.940	0.931	0.925
6000	0.961	0.945	0.928	0.917	0.908
8000	0.945	0.925	0.900	0.886	0.873
10000	0.931	0.902	0.872	0.853	0.840

Courtesy Compressed Air and Gas Institute

Not only will high altitude affect compressor operation, it will also affect electric motor operation.

Effect of Altitude on Electric Motors

Since the density of air decreases with increasing altitude, the ability of an air cooled electric motor to dissipate heat also decreases. Because of this, the power must be lowered. A chart is shown below.

Altitude (ft.)	Correction Factor
3000	1.0
4500	0.98
6000	0.95
8000	0.91
10000	0.88

Noise

Of ever increasing concern today is the noise a compressor generates. The Occupational Safety and Health Act (OSHA) covers many types of plant sound/noise conditions that not only may be disturbing to the plant personnel, but also damaging to their health.

Compressed air system sound arises from many different sources. One source of particular concern is the compressor installation. This will be examined in later sections. But what is acceptable sound or noise?

Acceptable Sound Levels

Impaired hearing may result after prolonged exposure to certain levels of sound/noise. However, there are many individual variances on what the ear can tolerate. The table below expressed the limitation excerpted from the Occupational Safety and Health Standards (1978).

Permissible Noise Exposures

Duration per Day in Hours	Sound Level dBA
8	90
6	92
4	95
3	97
2	100
1½	102
1	105
½	110
¼	115

Pinpointing Noise in the Compressor Room

The following are factors which may affect the sound levels nears the compressor:

1. Position of the compressor in the room, i.e., is it placed in a corner, along a wall, or in the center of the room?
2. The actual sound characteristics of the room, i.e., size of the room, the material used in the walls, ceiling and floor
3. Other equipment in the room that may generate, reflect or absorb sound
4. The mounting style of the unit – rigid or resilient?
5. Location of compressor inlet
6. The inlet and discharge piping lengths, along with whether they are flexible or rigid

If the sound is found to emanate from the unit itself, an enclosure may be built to contain the noise. Sound absorption material may be added to reduce the reverberating sound inside the box. Cooling is of utmost importance, and should be considered early in the enclosure design. Exiting ducts should be designed so as to minimize noise radiation to them. While some installations must be totally enclosed, others only need be shielded. A sound screen may be added to prevent noise from propagating in a particular direction. These barriers are made of sound isolating material, preferably lined with absorbing material on the compressor side. If there is a wall on the other side, an additional curtain should be added to reduce reflected noise.

Compressor House Ventilation

When discussing sound generated by a compressor, one remedy for excessive noise is compressor enclosure. However, in these cases, and even in standard compressor houses, ventilation is important. In most cases, heat dissipation through the ceiling, walls, windows and floor takes place, but is seldom sufficient. Typically, a fan is installed to move air through the building and carry away the heat. The amount of ventilation needed will, of course, depend upon the compressor installed. A chart is provided showing the amount of energy given off to the room as heat in percent of input shaft power.

Compressor Type	%
air cooled compressor	50
air cooled aftercoolers	45
electric motors	8
air cooled compressor with water cooled intercooler	3
water cooled compressor	2

Ventilation air should enter the room at a low point using short runs for ducts. Exiting air should be exhausted high at the opposite wall to avoid air stratification. Care must also be taken when positioning the exhaust so that the exhausting hot air and compressor noise do not interfere with nearby neighbors or operations.

Selecting a Compressor for a System

By now it should have become obvious that the selection and installation of a compressor system is no easy task. But a complete discussion of sizing a compressor to a system is beyond the scope of this text. However, a few important highlights are in order. Large compressor installations are expensive and complex. A good working relationship between the compressor specialists, manufacturers and plant operating personnel is needed when selecting a machine for a particular plant.

The compressor size selection depends upon the following points:

A. The system demand, including estimated base maximum loads, and near term forecast load additions

B. Standby capacity for emergency situations. A good way to handle this is to have an emergency connection piped to the system through the outside wall. With the appropriate shutoff and check valves, this can make it a simple matter to hook up a mobile unit in the event of a compressor failure.

C. The plant's future requirements.

Another consideration is, how many compressors are needed? A single unit is economical, but if it fails an entire plant is shut down. One single-stage compressor would usually be adequate for small systems up to about 100 scfm (47 dm³), it may be advantageous to use two or more compressors. System pressure of 65 to 80 psig (4.5 to 6 bar) can use single-stage compressors, while 100 psig (7 bar) systems are better served by two-stage units. The operating savings of a two-stage compressor as compared to a single-stage unit is about 17%.

Many times a battery of compressors is recommended for large systems or systems where demand is variable over a wide range. For example, consider a factory where the normal base load is 1000 scfm (472 dm³/s) with peak loads of 2000 scfm (944 dm³/s).

This system can be served by four compressors, each having a capacity of 500 scfm (236 dm³/s). Three of the compressors would normally operate 60 to 70% of their potential to meet normal service demands with the fourth on standby for peak loads and emergencies. Also, two 1000 scfm (472 dm³/s) units could be used. The two larger units could have a higher efficiency than the four

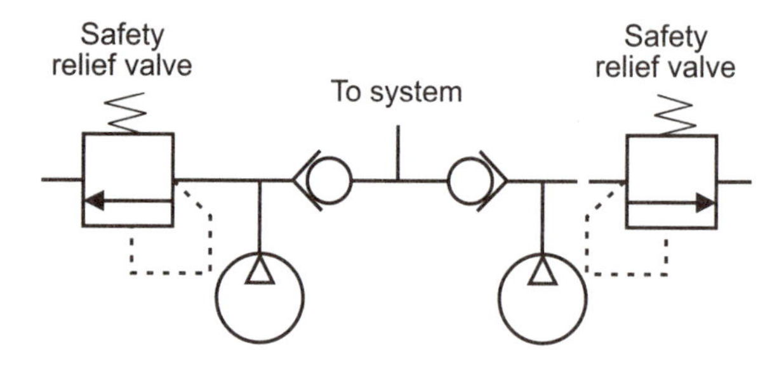

Figure 5-15

500 scfm (236 dm³/s) units. With the four-unit system, one standby helps to assure that the shop is never out of production because of a compressor failure. It should be noted that sophisticated controls would be needed for the four-compressor system.

Installation of a Compressor

But whether planning a new system or expanding an existing one, compressors should be placed close to the point of use. Savings are realized because of shorter supply lines, lower pressure drops and increased versatility.

When installing compressors, keep these rules in mind; compressor intakes should provide air as cool, dry and clean as possible, preferably by drawing the air from a dry, cool and clean air supply area outside the compressor house. Intakes should be shielded from the weather and kept away from dusty yards, parking lots, smokestacks and other discharge dusts and baffled to prevent compressor noise from being transmitted to the surrounding countryside.

The reason for taking air in from the outside is that, especially in the winter, it is cooler and therefore more dense. For instance, if the plant temperature is 80°F (26°C) the compressor takes in the same temperature air it delivers. But suppose the outside air is 50°F (10°C). For an output of 1000 ft³ (2832 dm³), the compressor will intake 943 ft³ (2671 dm³). This can be an appreciable savings in horsepower and therefore, in money. A chart is shown giving the correlation between intake temperature and relative delivery.

It should be noted that air inlets placed over black roofs may draw in very hot air on a sunny summer day or even on a winter day. This will happen especially if the air is relatively still. It may be a better idea to take air in along a high cool, shady wall or to paint the roof white or silver.

As for the size of the intake duct, the flow area should be about 25% larger than the piston area. If multiple units are used, the area of the duct at the compressor nearest the inlet should be equal to the sums of the piston areas of all the compressors. It is obvious, of course, that the line may be decreased in size as individual intakes are led off.

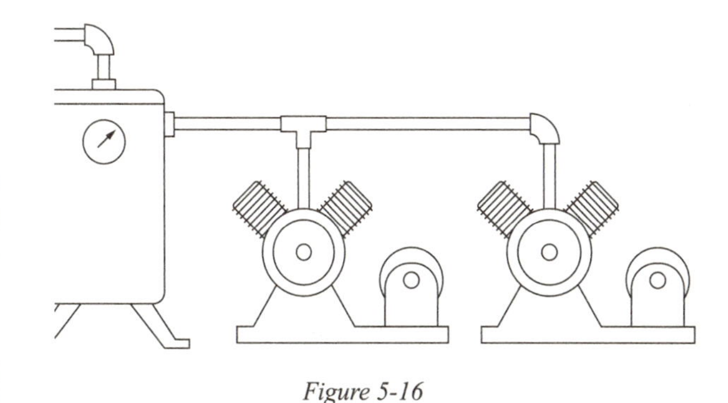

Figure 5-16

Effect of Initial or Intake Temperature on Delivery of Air Compressors
Based on a Normal Intake Temperature of 60°F

Initial Temperatures		Relative Delivery	Initial Temperatures		Relative Delivery
°F	°F abs		°F	°F abs	
-20	440	1.18	70	530	.980
-10	450	1.155	80	540	.961
0	450	1.13	90	550	.944
10	470	1.104	100	560	.928
20	480	1.083	110	570	.912
30	490	1.061	120	580	.896
32	492	1.058	130	590	.880
40	500	1.040	140	60	.866
50	510	1.020	150	610	.852
60	520	1.00	160	620	.838

The duct should be as short as possible. If a short duct is not possible, follow this rule: increase the duct one inch (25m) in diameter for every ten feet (3m) of length.

In addition, intake air should be filtered. Compressor manufacturers usually install the correct size and type of filter, including an intake silencer for the application. It is important to have regular filter maintenance as a filter is only as good as it is maintained. Clean filters allow compressors to do the job they are intended to do. Many times this filter gets dirty providing high intake resistance and is thrown away and not replaced. Remember, this filter is the first line of defense against dirt. For this reason, intake filters should be serviced at regular intervals to achieve optimum dependability.

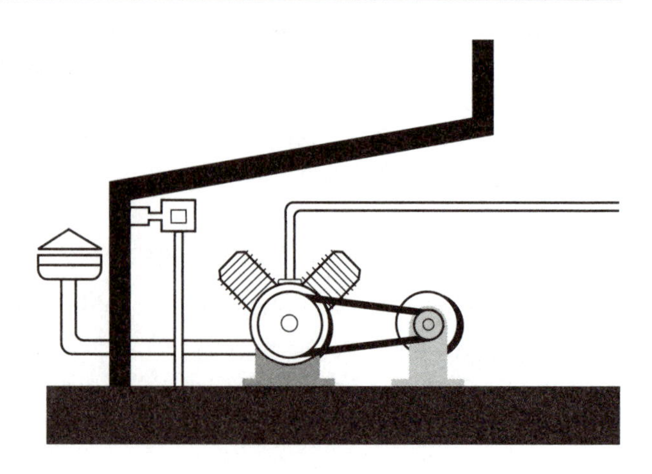

Figure 5-17

A pressure differential gage would aid in monitoring the pressure drop (condition) across the intake filter. This may be set up so that when the maximum pressure drop is reached a light is turned on or a horn is sounded, etc., to get the maintenance personnel's attention.

A dirty filter is like a great increase in altitude on the inlet side. It is difficult to measure the actual return on investment of the clean air filter. However, many surveys have been conducted, all of which show at least a 25% return. Much of the savings is realized in compressor cleaning charges and rebuild labor costs.

Approximate Compressor Tolerance to Airborne Particles

Compressor Type	Particle Size (Micrometers)
Reciprocating type (lubricated)	25
Reciprocating type (non-lubricated)	5 -10
Screw type (oil flooded)	10
Screw type (dry)	—
Centrifugal	1

Lesson review

- Compressors are available in two basic types: displacement and dynamic.
- Displacement compressors consist of piston, vane, screw and lobed rotor types.
- Centrifugal types consist of radial and axial compressors.
- If air requirements are such that a pressure greater than 100 psi (7 bar) is needed, a two-stage compressor should be employed.
- Because a compressor is sized for the total demands of a system, its output must be matched to that of the system. There are various compressor unloading techniques, the most common being holding the suction valve open.
- Increased altitude affects both the compressor and electric motor. Increased altitude means decreased performance.
- Noise in a compressor room is annoying and may be detrimental to personnel; techniques should be employed to decrease the sound pressure level.
- Ventilation is extremely important for a compressor house.
- Air filtration is necessary for a compressor. The manufacturer will supply the correct information for the correct type needed.

Exercise
Compressors
50 points

Instructions: Complete the statements below by filling in the blanks with the words of your choice. The correct answers can be found in the maze of letters below. Words run horizontally, vertically and diagonally.

T	V	B	L	T	O	N	E	N	N	I	F	I	N	I
S	P	R	D	W	S	I	R	E	H	K	D	R	Y	N
B	W	I	E	O	N	M	L	F	S	S	F	A	I	T
O	C	A	C	L	U	Z	H	Z	B	D	V	R	W	E
S	T	H	R	E	E	N	D	A	X	A	E	E	H	R
H	F	Y	E	H	R	S	M	Y	U	W	T	A	C	C
A	L	Q	A	T	P	O	T	E	N	T	I	A	L	O
T	O	Y	S	J	R	L	K	V	S	A	O	P	Z	O
A	O	G	E	W	S	L	O	O	P	H	M	E	I	L
G	D	P	S	O	D	N	D	W	C	M	K	I	G	E
S	E	A	E	H	I	G	H	C	L	P	P	G	C	R
K	D	C	E	N	T	R	I	F	U	G	A	L	I	N

1. Compressors convert mechanical energy transmitted by a prime mover into ____________ energy of compressed air.
2. A two-stage piston compressor has a(n) ____________ where compressed air is cooled while passing to a second stage.
3. Compressors may be displacement or ____________.
4. As altitude increases the output capacity of an air compressor ____________ .
5. According to OSHA, an unprotected person at work may occupy a room for ____________ hours at a sound level of 100 dBA.
6. When ventilating a compressor house, the inlet should enter ____________ while the outlet should exit ____________ to prevent stratification.
7. There are two types of helical compressors: ____________ and ____________ .
8. This compressor is a dynamic type; it is a ____________ compressor.
9. ____________ covers many types of plant sound/noise conditions.

Chapter 6
Aftercoolers, Driers, Receivers
Air Distribution Systems

Compressor Air

After being filtered to a large extent by the compressor intake filter, the air moves to the piston chambers. If it is a two-stage compressor, air is compressed with a consequent temperature rise in the first stage, passed through an intercooler where its temperature is decreased and then on to the second stage. In the second stage, the air is compressed further and has another increase in temperature.

The air at this point has increased in potential energy because it is compressed. But the air is hot and contains excess water vapor. In addition, air compressor outlet temperatures may run in the range of 400 to 500°F (204 to 260°C).

This air may contain some dirt which was not removed by the intake filter, and it may carry some oil vapor which was picked up while passing through the compressor. Both are undesirable. Heat is undesirable due to the fact that if this hot air is passed directly into the pipes, they would lengthen due to the heat. Contraction would occur upon compressor shutdown. This recurring process would cause joints to leak and reduce efficiency. Therefore, flexibility must be built in. Instead of discharging the air from the compressor outlet directly into an air receiver for storage, the air is passed through an aftercooler.

Aftercoolers

As its name implies, an aftercooler cools compressed air after compression has been completed. This is accomplished by passing cooling water or air over the aftercooler chamber.

Besides being the point where air cools, an aftercooler is also the place where some dirt and oil vapor fall out of suspension, and a good portion of entrained water vapor coalesces out. The aftercooler must have a moisture separator, preferably having an automatic drain.

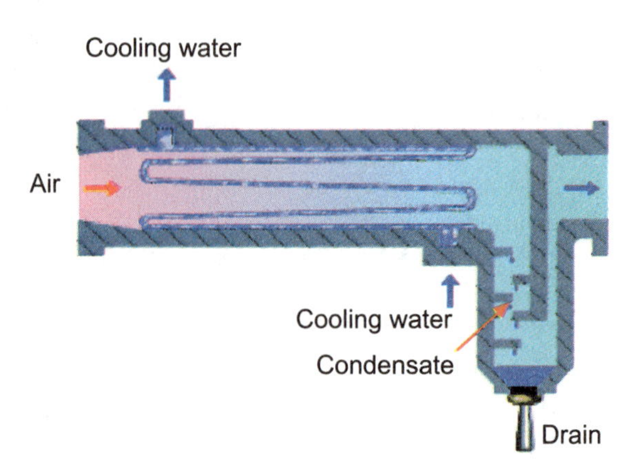

Figure 6-1 Aftercooler

How an Aftercooler Condenses Water Vapor

We know that when a gas cools, its specific volume decreases. This change usually results in a change (decrease) in the pressure of the gas. Also, when air cools, its ability to carry water vapor also decreases.

When a cool wind is felt on a humid summer's day, we know that it is probably going to rain. When hot air mixes with cooler air, the temperature decreases and so does the air's ability to hold water. Condensation results.

In an aftercooler, compressed air and the water vapor it carries is cooled and its water content is made to condense and to "rain". As air passes to their air receiver, it has much less potential energy than when it entered; it is cooler, cleaner and holds less water. The air leaving the compressor is typically very humid. This high humidity requires an aftercooler to remove some of the water vapor.

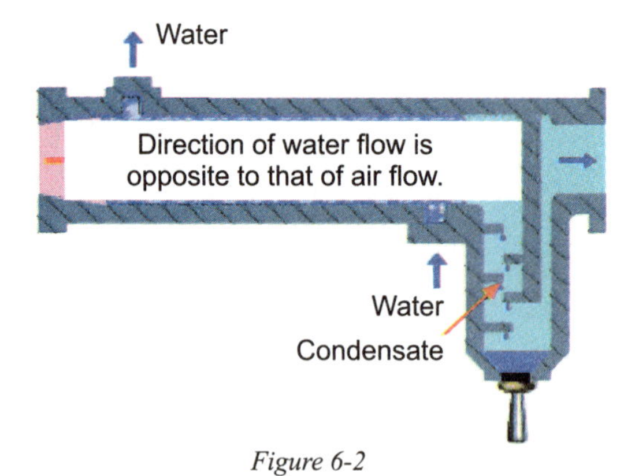

Figure 6-2

In a typical water type aftercooler, the direction of water flow is opposite that of the air flow. A good aftercooler will cool the air flowing through it to within 15°F of the cooling water temperature. It will also condense up to 90% of the water vapor originally contained in the air as it enters the receiver tank.

Water removal is as vital as heat removal. 1000 cubic feet (28 m^3) of air after compression can release as much as 1.4 quarts (1.3 L) of water. Thus a modest size, 100 scfm (472 dm^3/s) system could produce over 50 gallons of condensed water in a single 24-hour day. If demands of a system require drier air, the compressed air can then be put through the following processes:

1. over compression
2. refrigeration
3. absorption
4. adsorption
5. combination of the above methods

Overcompression

In the overcompression process, the air is compressed so that the partial pressure of the water vapor exceeds the saturation pressure. Then the air is allowed to expand, thereby becoming drier. This is the simplest method, but the power consumption is high. It is usually used for very small systems and therefore, is not as common in industry as are the other methods.

Refrigeration

As we mentioned before, when the temperature of air is lowered, its ability to hold gaseous water is reduced. This is what takes place in an aftercooler. However, the typical minimum air temperature attainable in an aftercooler is limited by the temperature of the cooling water or air. If extremely dry air is needed, a refrigerant type cooler is employed.

In these devices, hot incoming air is allowed to exchange heat with the cold outgoing air in a heat exchanger. The circuit is shown. The lowest temperature to which the air is cooled is 32.4°F (0.6°C) to prevent frost from forming. This type of air drying equipment has relatively low initial and operating costs.

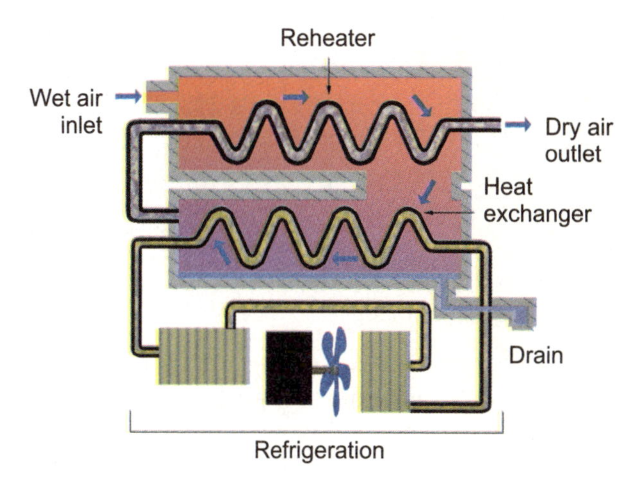

Figure 6-3

Absorption Process

Water vapor in compressed air can be removed from compressed air by methods such as absorption. There are typically two basic absorption methods. In the first, water vapor is absorbed in a solid block of chemicals without liquefying the solid. The chemicals used in the solid insoluble type are typically dehydrated chalk and magnesium perchlorate. Another type uses deliquescent drying agents like lithium chloride and calcium chloride, which react chemically with water vapor and liquefy as the absorption proceeds. Periodic replenishing must be performed.

Some problems tend to exist with the deliquescent drying process. In turns out that most of these drying agents are highly corrosive. Also, the desiccant pellets can soften and bake at temperatures exceeding 90°F (32°C).

This may cause an increased pressure drop. In addition, a fine corrosive mist may be carried downstream with the air and corrode system components. However, this type of air dryer has the lowest initial and operating cost of the more common air dryers. Maintenance is simple, requiring periodic replacement of the deliquescent drying agent.

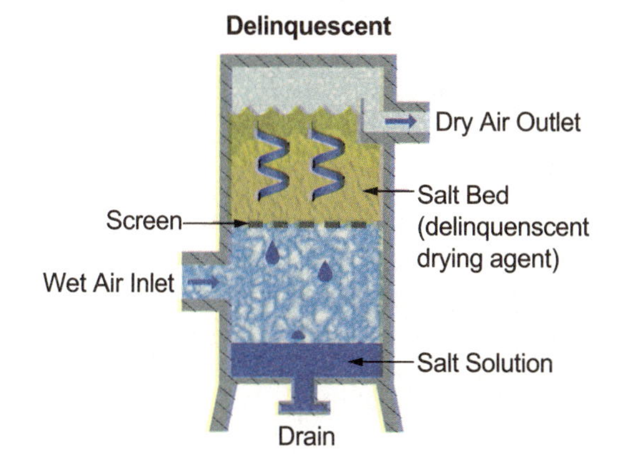

Figure 6-4

Adsorption

Adsorption (desiccant) drying is another industrial method for drying air. Adsorption chemicals hold water vapor in small pores in the desiccant chemicals. This action is not fully understood, and discussing it is beyond the scope of this text. Processes of this sort typically seek to make use of chemicals like silica gel ($Si\ 0_2$) or activated alumina (Al_20_3).

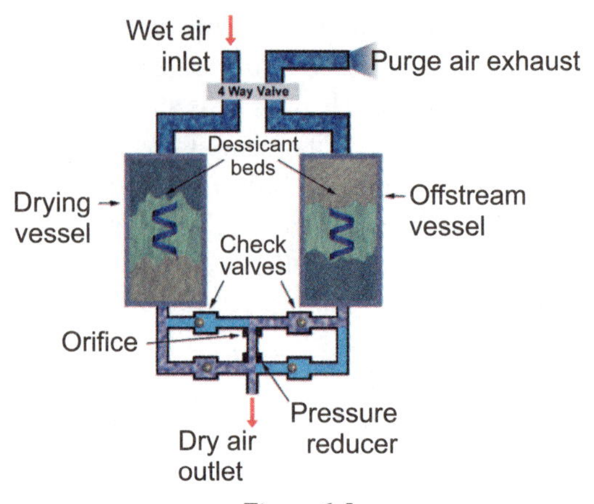

Figure 6-5

This type of air drying is the most costly of the drying methods discussed here. This is because of moderate to high initial and high operating costs. However, maintenance costs may be lower than the absorption type, because there are no moving parts. Also, the replacement of the desiccant is eliminated. But in a heat-regenerative type dryer, the maintenance cost may be higher due to the need to use high regenerative temperatures which may damage equipment.

Receiver Tank

After the air has left the aftercooler and/or dryer, it is directed to a receiver or storage tank. The receiver tank is used to store a supply of compressed air and assure a steady supply without excessive line pulsations or frequent loading and unloading of the compressor.

Figure 6-6 Receiver tank

A receiver tank is a pressure vessel for air storage, provided with a drain and flow ports and a flow baffle to assure optimum contaminant fallout and to provide ready access to devices which will drain off contaminants. Codes are quite strict as to the strength and testing of such devices. ASME (American Society of Mechanical Engineers), state and local codes require a safety valve be installed on most units. The valve must be capable of discharging the full rated flow of the compressor. Safety valves are sometimes set 10 psi (0.7 bar) above the normal receiver pressure, but always below the rated working pressure of the receiver. Always be sure the receiver has a drain, preferably an automatic one, at its lowest point. If the receiver is mounted horizontally, it is a good practice to tilt it slightly to ensure good drainage.

Sizing a Receiver Tank

Receiver tanks are sized according to the compressor output, size of the system and air demand cycles. It is better to oversize the receiver than to under size it. This adds very little to installation costs and by averaging varying flow demands, provides increased capacity to meet demand surges. It also lowers air velocity in the tank, which improves fallout of condensed and removal of droplets of solvents and oils which may be in the air stream.

One formula used for estimating the required receiver size is based upon compressed volume. It is:

$$V = \frac{Q \times Pa}{P_1 + Pa}$$

where: V = receiver size (cubic feet)
Q = compressor output (cubic feet per minute)
Pa = standard atmospheric (psig)
P_1 = compressed pressure (psia)

For US units: $V\ (ft^3) = \frac{Q\ (cfm) \times 14.7}{(P_1\ (bar) + 14.7)}$

For SI units: $V\ (m^3) = \frac{Q\ (dm^3) \times 0.60}{(P_1\ (bar) + 14.7)}$

However, because of system requirements, etc., a factor will be multiplied (1.5 to 3, selected according to experience) by the previous equation. The formula becomes:
For US units: $V = \frac{K \times Q \times 14.7}{(P_1 + 14.7)}$

Example:
A compressor has an output of 500 cfm (236 dm^3/s). The working pressure is 100 psig (7 bar).

What size receiver is needed?

Solution:
By experience, a K factor of 3 is selected. The formula becomes:

$$V = \frac{3 \times (500) \times 14.7}{(100 + 14.7)}$$

V = 192 ft^3 (5.4 m^3)

With a variable demand situation, another formula may be written: $V = \frac{t\ (Qout - Qin) \times Pa}{(P_1 - P_2)}$

Q out = flow rate out
Q in = flow rate in

where: V = receiver size
t = time to charge or discharge
Pa = atmospheric pressure
P_1 = initial working pressure
P_2 = final working pressure

For US units:

$$V(ft^3) = \frac{t\,(min)\,[Qout\,(cfm) - Qin\,(cfm)] \times 14.7}{[P_1\,(psi) - P_2(psi)]}$$

For SI units:

$$V(m^3) = \frac{t\,(s)\,[Qout\,(dm^3/s) - Qin\,(dm^3/s)]}{[P_1\,(bar) - P_2\,(bar)]}$$

The above formula is quite useful for sizing the receiver for varying flow demands.

Example:
The compressor has an output of 250 cfm and the system requires 350 cfm for 30 seconds. The initial working pressure is 100 psi. The pressure cannot drop below 80 psi. What size is needed?

Solution:

$$V = \frac{t \times [Qout - Qin \times 14.7]}{[P_1 - P_2]}$$

$$= \frac{(30/60) \times [350 - 250] \times 14.7}{20}$$

$$= 36.75\ ft^3\ (104\ m^3)$$

Also, a calculation can be made to find the recharging time.

Example:
How long will it take to recharge the receiver in the previous example if the demand decreases to 100 cfm?

$$V = \frac{t \times [Qout - Qin \times 14.7]}{[P_1 - P_2]}$$

Rearranging:

$$t = \frac{V\,[P_1 - P_2]}{[Qout - Qin] \times 14.7}$$

$$t = \frac{(36.75)\,(80 - 100)}{[100 - 250] \times 14.7}$$

$$t = 0.33\ min.$$

Care should be taken when selecting a receiver tank to see that it contains good baffling and that slow air flow exists. This is what allows entrained contaminants (liquids and gases) to fall to the bottom of the receiver. Also, local codes should be strictly followed.

Piping Systems

Now, once the air has been compressed, cooled and dried by the aftercooler and stored in the receiver tank, it is ready for the piping system. The three main types of piping systems most commonly used are:

1. dead end or grid
2. unit or decentralized
3. loop

Grid Systems

The grid system is the simplest of the piping systems. It is often called the dead end system. It is simple in construction, consisting of a central main with small feeder lines and headers. The mains decrease in size away from the compressor, while feeder lines are generally of uniform size. Outlets are provided at convenient points on the feeder lines and points between feeders may be served by cross connecting any two adjacent legs.

While perhaps this is the simplest system and the least expensive to install, only one flow path is available and work stations near the ends of the system are subject to insufficient air supply (air starvation) when upstream demand is heavy.

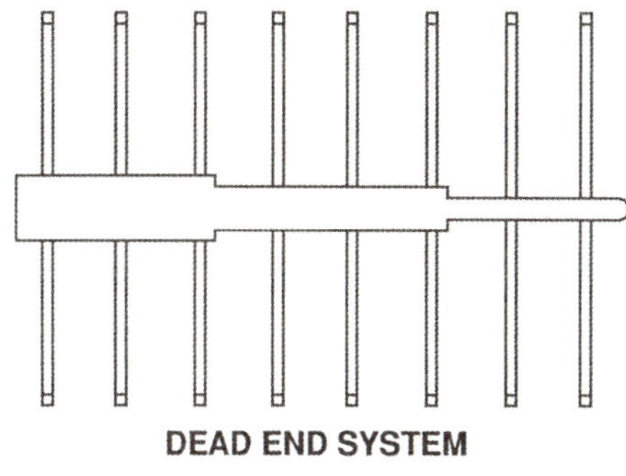

DEAD END SYSTEM

Figure 6-7

Unit or Decentralized Systems

The next unit or decentralized system may consist of two or more grids, each with its own compressor. The individual units may be interconnected if desired. Compressors are closer to the system using the air, allowing shorter supply lines. This means lower pressure drops, resulting in a more uniform air supply and system pressure. Decentralized systems are more versatile than single grids and more easily adapted to changing requirements.

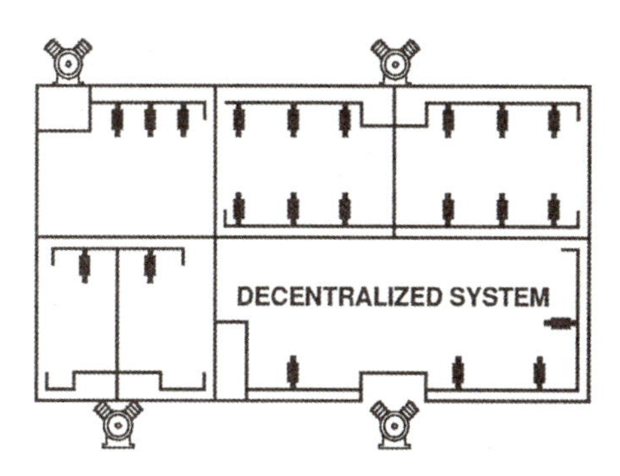

Figure 6-8

Loop Systems

Lastly, the preferred or recommended system is the loop. This type allows the optimum conductor size and assures more equal distribution through the plant. This arrangement provides a parallel path to all work points. At points of heavy momentary demands for air, a receiver can be used to store the energy for peak demands, preventing serious pressure loss and air starvation.

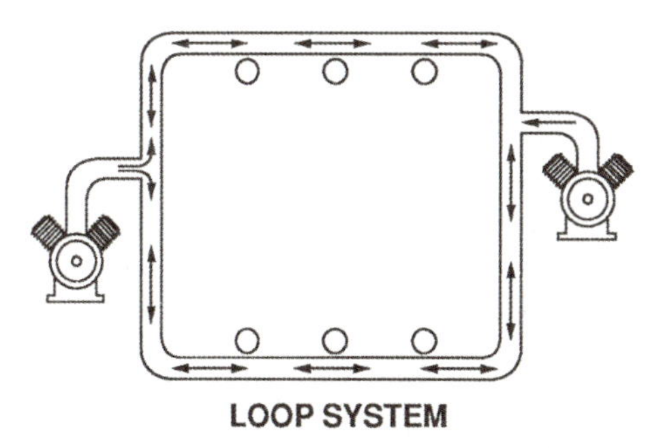

LOOP SYSTEM

Figure 6-9

Installation Consideration

However, when laying out the circuit or distribution loop, give careful attention to reducing the number of fittings to a minimum and keeping bend radii as large as possible. Tubing, because it is smoother than pipe, provides better flow and reduces pressure drop. Its real advantages are the major reduction in the number of fittings required with fewer possible leak points and smaller pressure drop. Pipe fittings generally have five times the pressure drop of an equivalent size 90° bend.

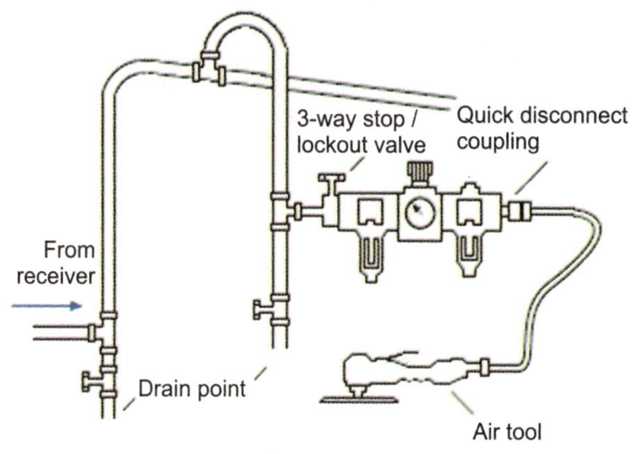

Figure 6-10

Distribution mains, feeders and headers should have a small slope (perhaps as much as ¼" per foot) to insure that condensed water will be swept by air flow to drains. Water legs fitted with automatic drains should be provided at low spots on the main and on drop legs. Drop legs should be taken from the top of the main header, with the bend made of the same size conductor as the leader line, or with a bend of 180° around a radius large enough to ensure low pressure drops. By taking air from the top of the main, most dirt and water are excluded. Drains, preferably automatic, should be provided at the bottom of each drop line at the end of a water leg.

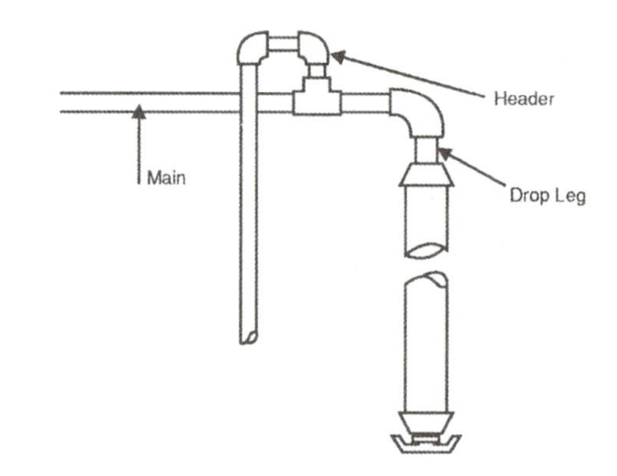

Figure 6-11

High Cost of Air Leaks

Although pneumatic leaks do not usually promote a housekeeping problem like hydraulic leaks, fixing them is still important. In the typical system, as we have described in this section, checking for leaks is the important final step. Leaks waste air and reduce air-tool performance.

In terms of air cost alone, leaks equivalent to a ½" (13 mm) hole can waste about 12 million cubic feet (34000 m^3) of air per month. At an average cost of 10 cents per 1,000 cubic feet (472 dm^3/s) of compressed air, this leakage costs about $1,200 per month. The major cost of leakage is the cost of electricity to keep the compressor running to make up the losses. Of course, the longer the compressor runs, the more maintenance is required, and finally the life of the compressor is reduced.

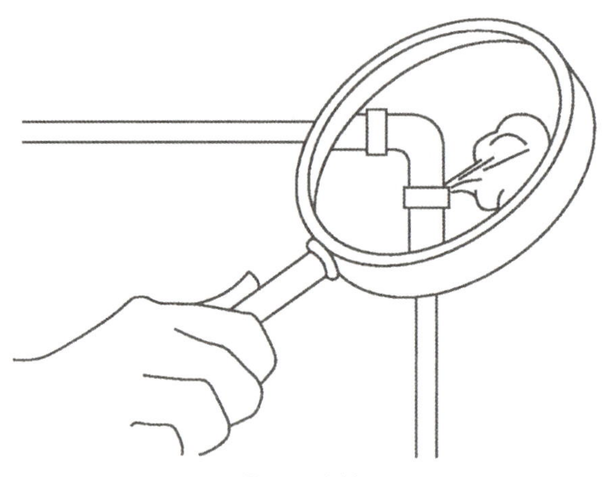

Figure 6-12

A leaky air system may drastically lower air-tool efficiency by lowering both pressure and air flow available to the tool. For instance, in some cases a pressure drop to just 10 psi (0.7 bar) can result in a drop in tool efficiency of 15%. The lowering of tool performance becomes disproportionately worse as pressure at the tool is reduced. Therefore, it is obvious that we must keep the leaks to a minimum by checking the system for leaks.

Check for Leaks

Frequently, serious leaks are audible. It is a good idea to "listen" for leaks when most of the plant is shut down for a shift change or at the end of a work week. As another test, soapy water solutions or commercial leak detecting fluid can be applied to suspected runs or connections. One very effective method of automatically detecting the presence of leaks and their order of magnitude is the two clock system. The clocks are connected to the compressor system by automatic controls. One clock shows total elapsed time. The other is circuited through the motor control and runs only when the compressed is under load.

Then at predetermined times, such as on weekends or at night when the air system is not in use, the clocks are started and run for a set period, usually one hour, and then stopped. Comparing the clocks makes it possible to determine the amount of air lost through leaks. Automatic equipment can make a permanent record.

Still another and perhaps faster procedure is to shut the system down and time how long it takes for system pressure to drop to a minimum value. This could be done at night when the plant is down. It is during this period that there is time to repair any leaks and problems.

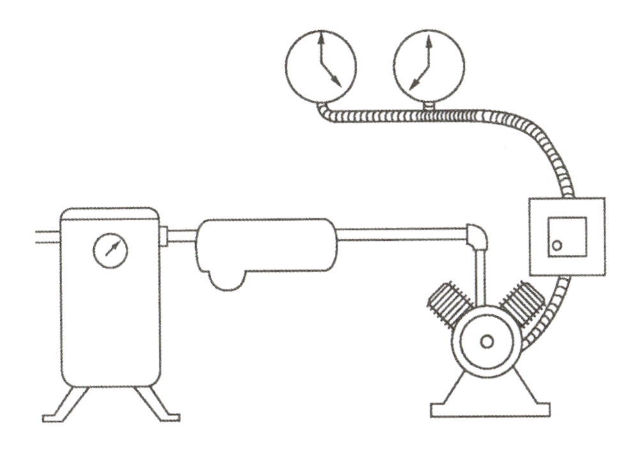

Figure 6-13

Lesson Review

- After air is compressed, it has an increased pressure and temperature. It must be cooled. An aftercooler will be used to decrease the air's temperature and reduce the water and oil in the air.
- If extremely dry air is needed, dryers may be used.
- Air leaving the aftercooler and/or dryer must be placed in a receiver tank. This is a place where air is stored while being further cooled, allowing additional water to precipitate. Size a receiver according to the guidelines provided in this chapter.
- Three types of piping systems exist: grid, decentralized and loop.
- Leaks in a compressed air system decrease overall efficiency.

Exercise
Aftercoolers, Driers, Receivers Air Distribution Systems
50 points

Instructions: Complete the statements below by filling in the blanks with the words of your choice. The correct answers can be found in the maze of letters below. Words run horizontally, vertically and diagonally.

A D E C E N T R A L I Z E D R
B P R R K S L R E I K I O R E
S W J E A N O S F S S F A I F
O C A C L U C H Z T D V R W R
R T H E E E A D A E A E E H I
P B Y I H R L M R N W T A C G
T D Q V T L E A K Y A S M E E
I D Y E J R L K V S E I K Z R
O M G R W S L O O P H R E I A
N F P T O D N D W C M J P G N
S A D S O R P T I O N P G O T
K T W O A F T E R C O O L E R

1. If extremely dry air is needed, a ◯_ _ _ _ _ _ _ _ _ _ _ _ _ _ type dryer is employed.
2. After air has exited the aftercooler and/or _ _ _ ◯ _, it is directed to a _ _ _ _ ◯ _ _ _ .
3. Which piping system may consist of two or more grids, each with its own compressor?
   _ _ _ _ _ ◯ _ _ _ _ _ _ _ _
4. What drastically lowers air tool efficiency by lowering both pressure and air flow available to the tool?
   _ _ _ ◯ _ air system
5. One good way to check for air leaks is to just _ ◯ _ _ _ ◯ for them.
6. _ _ _ ◯, state and local codes require that a safety valve be installed on receiver tanks.
7. The three basic piping systems are: grid, decentralized and ◯ _ _ _ .
8. A(n) _ ◯ _ _ _ _ _ _ _ _ _ _ _ _ causes air to cool and reduces its water content.
9. It's the process where water vapor is held by molecular surface tension. ◯ _ _ _ _ _ _ ◯ _ _ _

Question:

A compressor's first line of defense against contamination is its ◯◯◯◯◯◯ ◯◯◯◯◯◯

Instructions: Solve the following problem.

10. A compressor has an output of 100 cfm (482 dm^3/s). A maximum demand of 1300 cfm (614 dm^3/s) is needed for 20 seconds. The initial working pressure is 120 psi (9 bar). The pressure cannot drop below 90 psi (6 bar). What size is needed?

Chapter 7
Check Valve, Cylinders and Motors

In this lesson we will concentrate on check valves and the fluid power actuators of cylinders and motors. A basic check valve is a component which allows flow in only one direction.

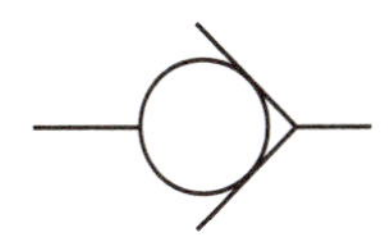

Figure 7-1 Check valve symbol

The check valve basically consists of a valve body with inlet and outlet ports and a moveable member which is lightly biased by spring pressure. The moveable member can be a flapper or plunger, but most often it is a ball or poppet.

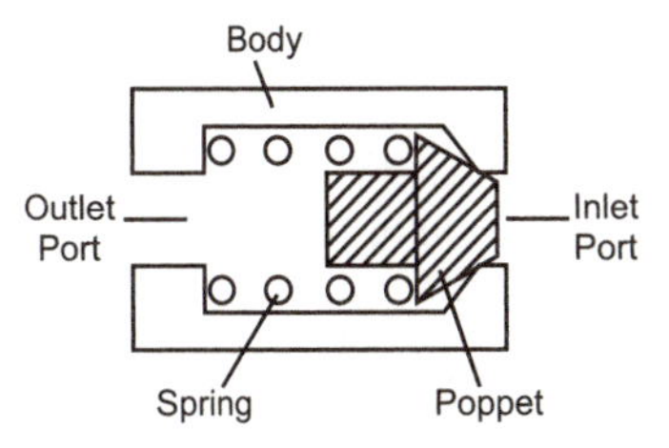

Figure 7-2 Check valve components

How a Check Valve Works

Because of its construction, air flow may pass through a check valve in one direction only.

When system pressure at the check valve inlet is high enough to overcome the low spring force, (5 psi) (0.4 bar) biasing the poppet against its seat, the poppet is moved off its seat. Flow passes through the valve. This is defined as the check valve's free flow direction. When air flow attempts to enter through the outlet, the poppet is pushed on its seat. Flow through the valve is virtually blocked.

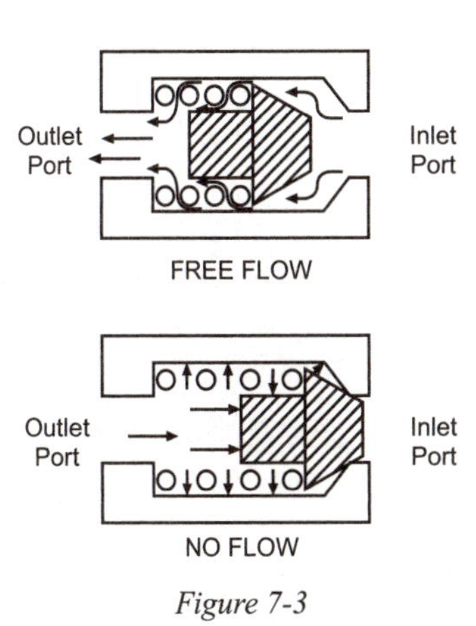

Figure 7-3

Check valves in pneumatic systems generally use a resilient seal on the poppet. A check valve is often used in systems as a bypass valve. It allows flow to get around components, like flow control valves which otherwise restrict flow in both directions.

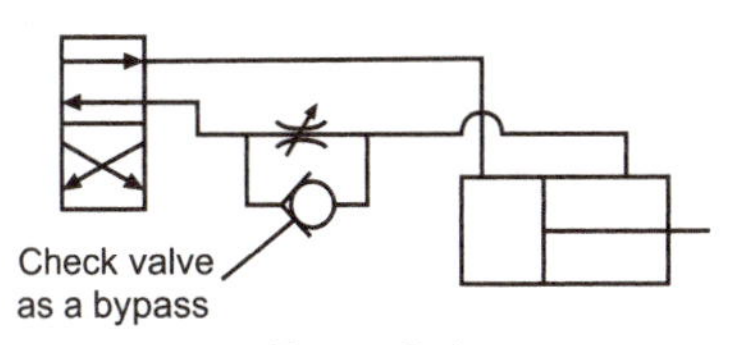

Figure 7-4

Cylinders

The most common pneumatic actuator is the cylinder. In all applications, pneumatic energy must be converted to mechanical energy for useful work to be done. The cylinder converts fluid power energy into straight-line mechanical energy.

The pneumatic cylinder consists of a cylinder body, a moveable piston, and a piston rod attached to the piston. End caps may be attached to the cylinder body barrel by many methods - threads, keeper rings, tie rods or welds. (Industrial cylinders often use tie rods.)

As the piston rod moves in and out, it is guided and supported by a removable bushing and seal assembly called a rod gland. The side through which the rod protrudes is called the "head." The opposite side without the rod is termed the "cap". Inlet and outlet ports are usually located in the head and cap.

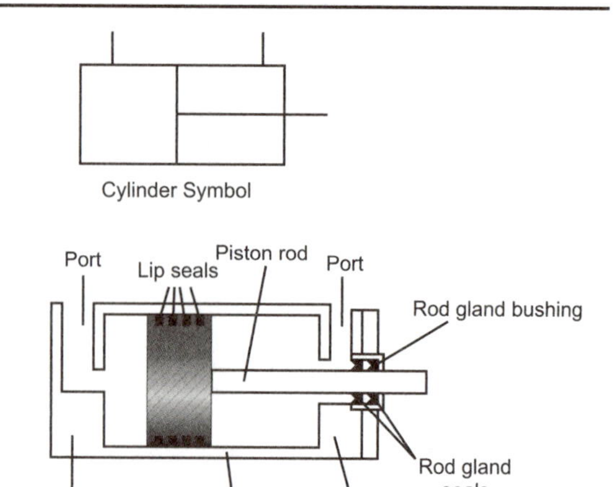

Figure 7-5

Seals

For proper operation of the pneumatic cylinder, a positive seal must exist across a cylinder's piston as well as at the rod gland.

Pneumatic cylinders usually are provided with a resilient piston seal. Resilient seals do not leak under normal conditions, but are less durable than either metallic seals or O-ring, lip seals, leather cups, etc.

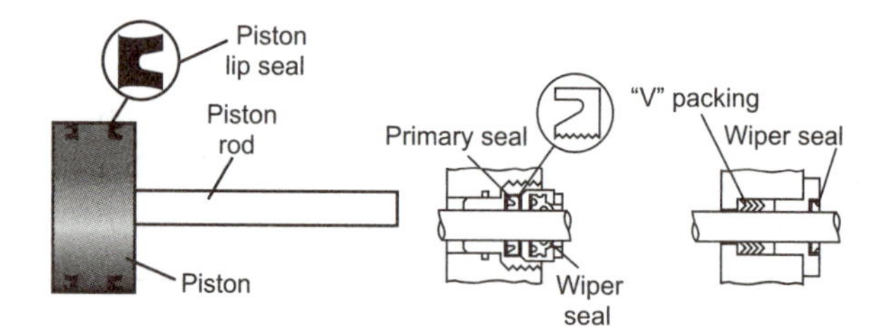

Figure 7-6

Rod gland seals also come in several varieties and are generally resilient seals. Some cylinders are equipped with a U, V, multi-lip or cupped shape primary seal and a rod wiper which controls the ingression of foreign materials at the rod gland.

One popular type of rod gland seal has a lip seal as its primary seal. The edge of the lips contacts the cylinder rod continuously to give a positive seal at low static and dynamic forces. This is used in conjunction with a wiper seal which collects any fluid which manages to pass the primary seal during rod extension and wipes the rod clean during rod retraction and then deposits a thin film of oil on the rod to be returned to the cylinder on its return stroke.

Stroke Adjusters

Sometimes the stroke of a cylinder must be externally adjusted to specific limits. Adjustment may be accomplished with a threaded rod which can be screwed as a stop for the piston when it is in the retract mode.

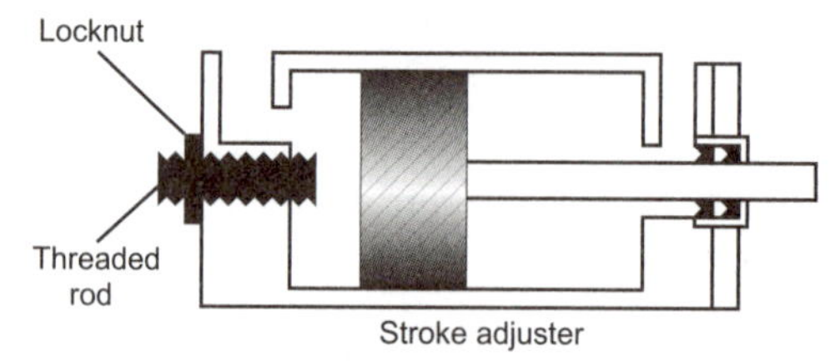

Figure 7-7

Cylinder Mounting Styles

Cylinders can be mounted in a variety of ways. Commonly used mounts are flange, trunnion, side lug and side tapped, clevis, tie rod, bolt mounting and combinations of these.

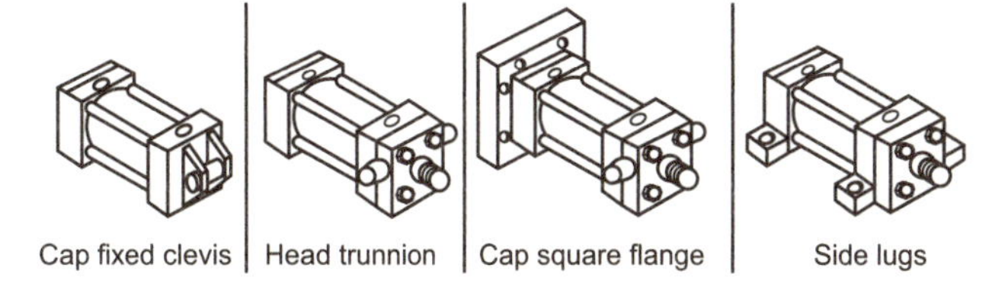

Figure 7-8

Mechanical Motions

As stated before, cylinders convert energy into straight line or linear mechanical motion. But depending upon the way in which they are attached to mechanical linkages, cylinders may provide a variety of non-linear mechanical motions.

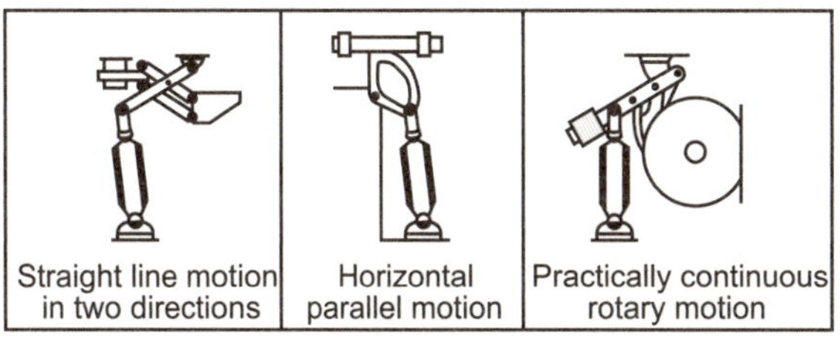

Figure 7-9

Types of Cylinder Loads

Cylinders can be used in a large number of applications to move various types of loads. A load which is pushed by a cylinder rod is termed a thrust or compressive load, while a load which is pulled by a cylinder rod is called a tension load.

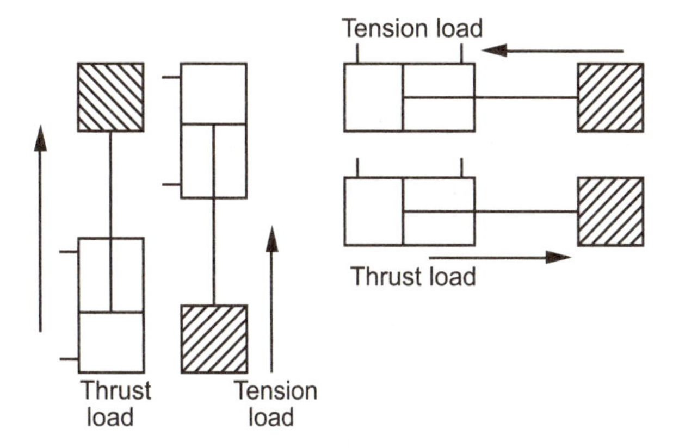

Figure 7-10

Common Types of Cylinders

There are many different cylinder types. The most common are discussed here.

- Single-acting cylinder - a cylinder in which air pressure is applied to the moveable element in only one direction.
- Spring return cylinder - a cylinder in which a spring returns the piston assembly.
- Ram cylinder - a cylinder in which the moveable element is the piston rod.
- Double-acting cylinder - a cylinder in which air pressure may be alternately applied to a piston attached to a moveable element to drive it in either direction.
- Single rod cylinder - a cylinder with a piston rod extending from one end.
- Double rod cylinder - a cylinder with a single piston and a piston rod extending from each end. The rods may be significantly different diameters.

As you can see, there are many types, styles and arrangements of cylinders. But unless they are sized correctly, they will not work effectively.

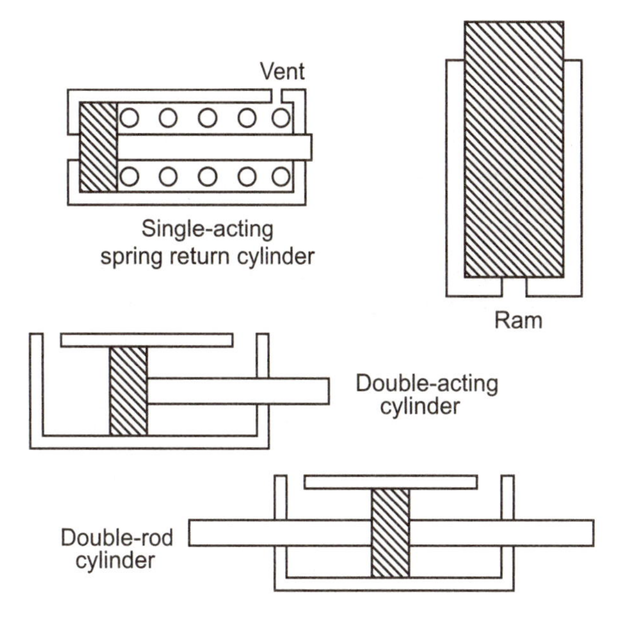

Figure 7-11

Sizing a Cylinder

To determine the size cylinder that is needed for a particular system, certain parameters must be known. First of all, a total evaluation of the load must be made. This total load is not only the basic load that must be moved, but also includes any friction and the force needed to accelerate the load. Also included must be the force needed to exhaust the air from the other end of the cylinder through the attached lines, control valves, etc. Any other force that must be overcome must also be considered a part of the total load. Once the load and required force characteristics are determined, a working pressure should be assumed. This working pressure that is selected **MUST** be the pressure seen at the cylinder's piston when motion is taking place. It is obvious that the cylinder's working pressure is less than the actual system pressure due to the flow losses in lines and valves.

With the total load (including friction) and working pressure determined, the cylinder size may be calculated using Pascal's Law. Force is equal to pressure being applied to a particular area. The formula describing this action is:

FORCE = PRESSURE X AREA

In U.S. Units: F (#) = P (psi) x A (in^2)

In S.I. Units: $\dfrac{P\ (bar) \times A\ (cm^2)}{0.1}$

Let us now look at a cylinder with no motion. In order to derive the complete force equation for a cylinder with no motion, consider Figure 7-12. This cylinder has a pressure Pc acting on the piston area. A pressure, Pr, acts on the minor area. "F", which is an external force, keeps the rod in compression. To find the force developed by Pc, we must find the area of the piston.

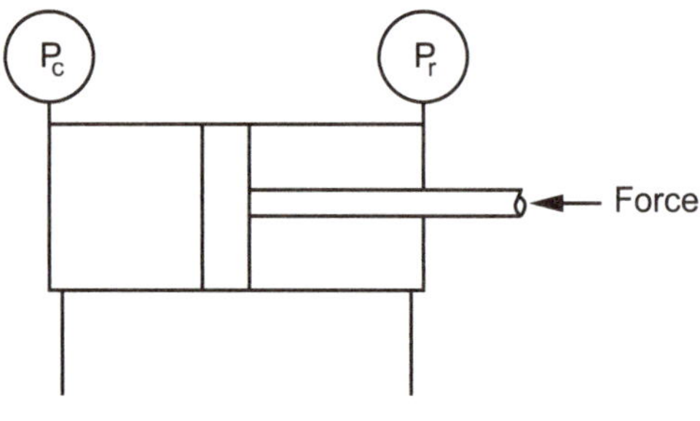

Figure 7-12

$$Ap = \frac{\pi\, dp^2}{4}$$

The area of a circle (in this case the piston and its seal) equals (3.14) or π times the diameter squared all divided by 4 or:

$$Ap = \text{Piston Area} = \frac{dp^2}{4} = \frac{3.14\ dp^2}{4} = 0.7854\ dp^2$$

Where dp = diameter of piston (in).

When the pressure Pr acts on the minor area Ae, the minor area is the area of piston minus the area of the rod. Consider the figure at the right. Pneumatic pressure acts on the effective area, but it cannot act on the area covered by the rod. This means that the new are we need to find is Ae, or the area of the piston minus the area of the rod.

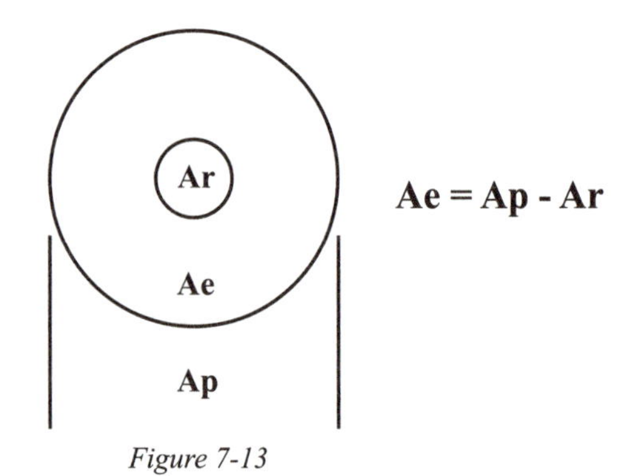

Figure 7-13

For S.I. Units

$Ae\ (in^2) = (Ap\text{-}Ar)\ \frac{\pi}{4}\ (dp^2 - dr^2) = 0.7854\ (dp^2 - dr^2)$

For S.I. Units

$Ae\ (cm^2) = 0.007854\ [dp^2\ (cm) - dr^2\ (cm)]$

Where

Ap = piston area
Ar = rod area
dp = diameter of piston
dr = diameter of rod

Therefore, the equation for a compressive load becomes:

Fc = (Pc x Ap) - (Pr x Ae)

Where:

Fc = the load in compression
Pc and Pr are the pressure at the cap and load end respectively.
Ap and Ae are the major and minor areas

If the load is tensile (or the rod is pulling on the load), the formula will become:

Ft = (Pr x Ae) - (Pc x Ap)

For the purpose of simplicity we shall assume static (including seal) friction to be zero. However, it should be noted that this could be a fairly significant mount.

Example 7-1

A designer is asked to select a cylinder which will push an equivalent load of 250 pounds (1111 N). The shop supply air is at 100 psi (7 bar). What size cylinder is needed?

Solution:

Since the load is known, we need to assume a pressure. Since the shop air is 100 psi, a value of 85 psi or less would be logical. But for this problem let us select 50 psi to work with. Since the load is in compression, we will apply the formula- -

Fc = (Pc x Ap) - (Pr x Ae)

Since the pressure Pr is close to zero, we will assume it to be that value. This arranges the equation to read:

$Fc = Pc \times Ap$ ***or*** $Ap = \frac{Fc}{Pc}$

Substituting:

$$Ap = \frac{250}{50} = 5 in^2 \ (32.2 \ cm^2)$$

Using formula for area and rearranging it:

$$dp = \sqrt{\frac{Ap \times 4}{3.14}} = \sqrt{\frac{5 \times 4}{3.14}}$$

$$dp = \sqrt{6.36} = 2.52 \text{ in. } (64 \text{ mm})$$

An alternate method, once the piston area is known, is to refer to force charts in cylinder manufacturers' catalogs to convert piston area into piston diameter. Always select the next larger piston diameter.

To select a standard size bore, a 3.25 inch (82 mm) would be the choice, or a 2.5 inch (63 mm) cylinder running at 51 psi (3.5 bar) again assuming friction and acceleration forces to be zero.

Example 7-2

Select a cylinder to lift one ton (8,889N) using a working pressure of 60 psi (4 bar).

Solution:
Since the rod is in tension the formula below will be used.

$$Ft = (Pr \times Ae) - (Pc \times Ap)$$

Assuming Pc = 0, then:

$$Ft = Pr \times Ae \quad \textit{or} \quad Ae = \frac{Ft}{Pr}$$

Substituting:

$$Ae = \frac{2000}{60} = 33.3 \ in^2 \ (214.8 \ cm^2)$$

Applying equation (6 - 3a)

$$Ae = Ap - Ar$$

By a series of calculations, we find that a standard cylinder with an 8" bore (200 mm) and a 1-3/8" (35 mm) rod will work.

Stop Tube

Now that we have a bore and rod diameter, we must examine the cylinder for internal high bearing stresses. To correct for this and extend the life of the cylinder and the rod gland, we would use a stop tube.

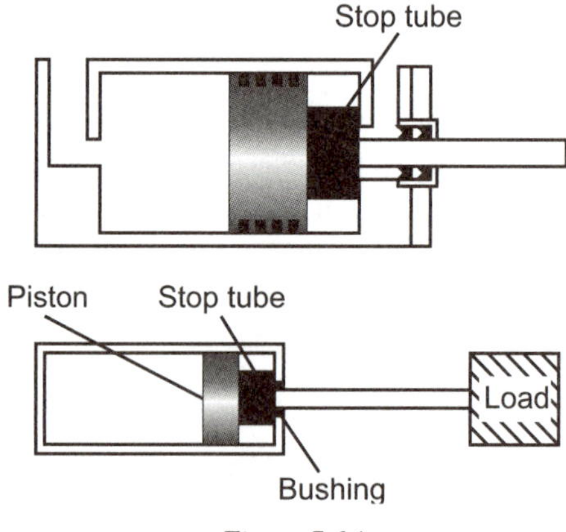

Figure 7-14

A stop tube is a metal collar which fits over the piston rod. It keeps the piston and rod gland bushing separated when a long stroke cylinder is fully extended.

Since it is a bearing, a rod gland bushing is designed to take some side loading when supporting the rod as it extends or retracts. Along with being a bearing, a rod gland bushing is also a fulcrum for the piston rod. A stop tube in effect decreases the rod gland bushing loading at full extension between both piston and bushing.

Steel rods of long stroke cylinders sag because of their weight. A 5/8" (16 mm) diameter piston rod weighs 1 lb. per foot (1.5 kg./m) and will sag over 1 in. (25 mm) at the center of a 10 ft. (3.05 m) span. A stop tube is used to separate the bushing and piston when the rod is extended. This in effect reduces the load on the rod gland bushing by moving the fulcrum point.

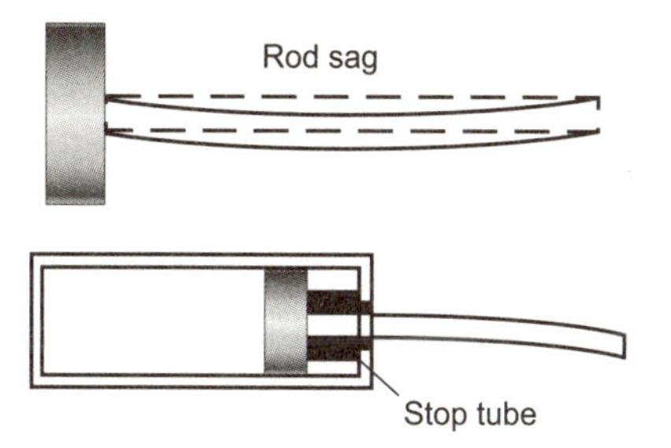

Figure 7-15

Selecting a Stop Tube

Selecting a stop tube depends upon the basic length and the maximum thrust developed. The basic length is a function of the actual stroke of the cylinder and the type of cylinder mounting.

The formula for basic length is:

Basic length = stroke x stroke factor

Stroke factor can be determined from (Fig. 7-17). As seen from the figure, the stroke factor increases as the cylinder becomes less rigid. Case 6, pivoted and rigidly guided, is four times worse than fixed and rigidly guided.

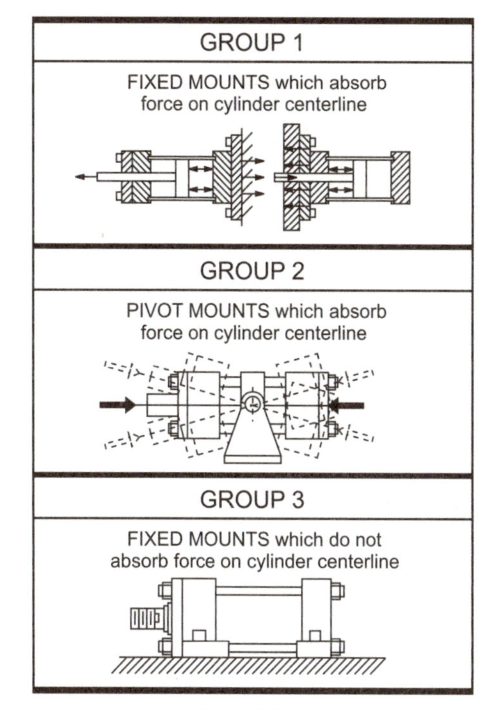

Figure 7-16

RECOMMENDED MOUNTING STYLES FOR MAXIMUM STROKE AND THRUST LOADS	ROD END CONNECTION	CASE	STROKE FACTOR
GROUPS 1 OR 3 Long stroke cylinders for thrust loads should be mounted using a heavy-duty mounting style at one end, firmly fixed and aligned to take the principle force. Additional mounting should be specified at the opposite end, which should be used for alignment and support. An intermediate support may also be desirable for long stroke cylinders mounted horizontally. See Catalog under "Tie Rod Supports – Rigidity of Envelope" for a guide. Machine mounting pads can be adjustable for support mountings to achieve proper alignment.	FIXED AND RIGIDLY GUIDED	I	.50
	PIVOTED AND RIGIDLY GUIDED	II	.70
	SUPPORTED BUT NOT RIGIDLY GUIDED	III	2.00
GROUP 2 Style D – Trunnion on Head	PIVOTED AND RIGIDLY GUIDED	IV	1.00
Style DD – Intermediate Trunnion	PIVOTED AND RIGIDLY GUIDED	V	1.50
Style DB – Trunnion on Cap or Style BB – Clevis on Cap	PIVOTED AND RIGIDLY GUIDED	VI	2.00

Figure 7-17

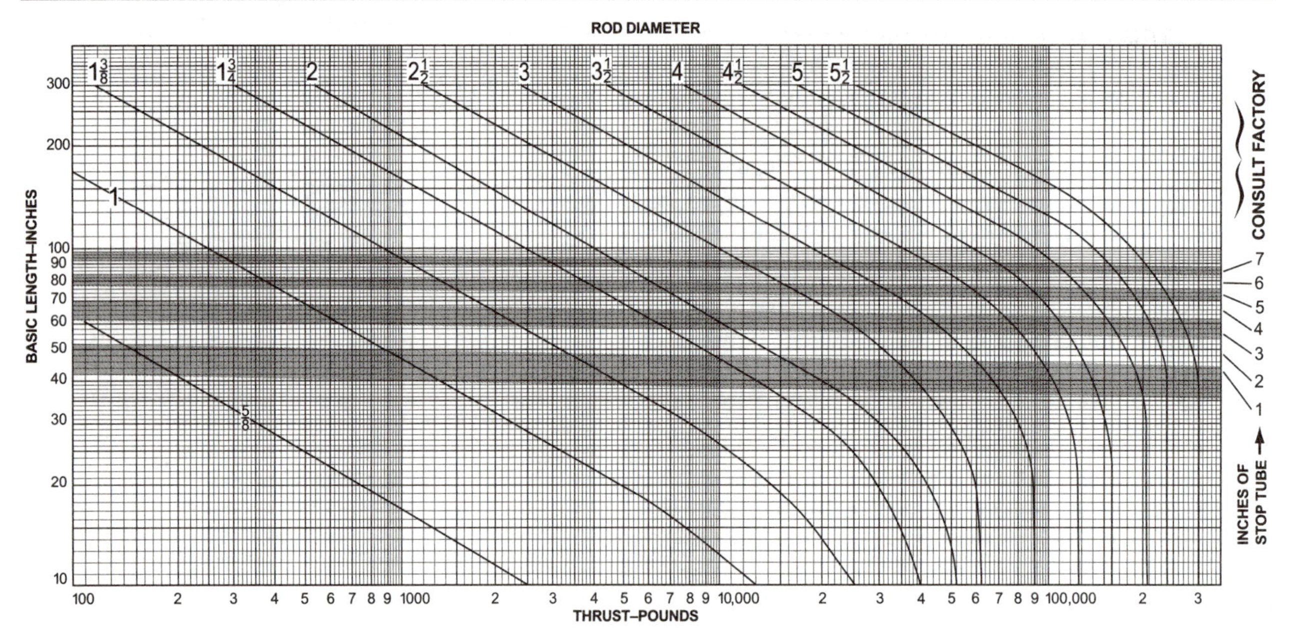

Figure 7-18

After the basic length is found, (Fig. 7-18) may be used. Thrust is plotted vs. basic length. The stop tube is found in the center of the graph and varies from 1 to 7 inches for a basic length of 35 to 100 inches.

Where thrust and basic length intersect will determine the stop tube length.

Once the need for a stop tube has been checked, a rod diameter must be selected.

Buckling

Cylinders may fail by buckling at a thrust load which has induced rod stresses far less than the elastic limit of the material. Buckling is the sudden collapse of the rod at or above the critical thrust load.

Failure of cylinders due to buckling is dependent upon several factors:

1. The axial loading; i.e. the resulting thrust forces on the rod.
2. The bending moment imposed on the rod by the load - due to eccentric loading on torque.
3. The type of end conditions that exist on the cylinder, i.e. cylinder mounting style, external transverse forces on the cylinder.

The type of end conditions on the cylinder was discussed in the previous section on selecting a stop tube.

Sizing a Rod to Prevent Buckling

After the basic stroke length of the cylinder is determined, (see **Selecting a Stop Tube**) Fig. 7-18 is applied for the minimum rod size for reassembly centered thrust load forces. The thrust and basic stroke length are intersected and the next largest rod is selected (rod line seen at a 45° angle left to right).

Example 7-3

In a particular application, a designer uses a cylinder to develop 1000 lb. (4445 N) of thrust. You mention to him that for a stroke of 32" (813 mm) with the cylinder trunnion mounted at the cap end and the load pivoted and rigidly guided, he must have a stop tube and at least a 1-3/8" (35 mm) diameter rod. He says you're wrong. Prove who is correct.

Solution:

To determine whether a stop tube is needed or buckling will take place, the basic length must be calculated first.

Basic length = stroke x stroke factor

For the problem given, the stroke factor is a 2.

Basic length = 32 x 2 = 64 inches (1625 mm)

Entering Fig. 7-3 with a basic length of 64" and a thrust factor of 1000, we can find the rod diameter and stop tube length.

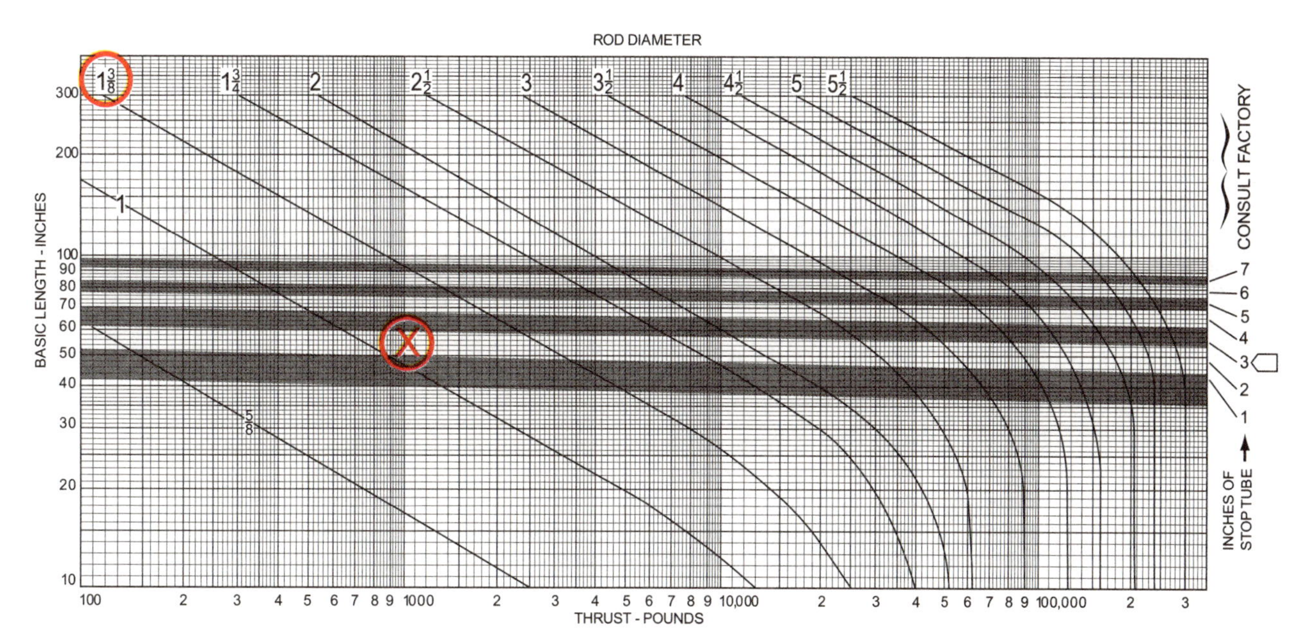

Figure 7-19

A stop tube length of 4" (76 mm) is found at the right hand side of the graph. The rod diameter is 1-3/8" (35 mm) and is found by the first rod size moving vertically upward from the point of intersection. Now we must examine the need for cylinder cushions.

Cushions

When pneumatic energy moving a cylinder's piston is suddenly stopped (as at the end of a cylinder's stroke) , the kinetic energy is changed into a concussion known as "shock". If a substantial amount of kinetic energy is stopped too abruptly, the resulting excess shock may damage the cylinder .

To help protect against excessive shock, a cylinder can often be equipped with cushions which will absorb the kinetic energy of the moving system. Cushions slow down a cylinder's piston movement just before reaching the end of the stroke. Cushions can be applied at either or both ends of a cylinder. However, standard cushions have a definite limit on the amount of kinetic energy which they will safely absorb. Beyond these levels, other measures must be taken to control the load and its kinetic energy, such as cam, solenoid or pilot-operated deceleration circuits or the use of external absorbers.

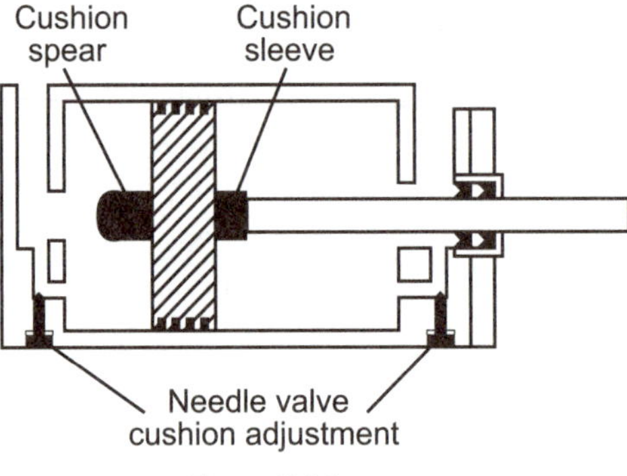

Figure 7-20

A cushion consists of a plug attached to a piston, which will trap a volume of air and compress it as the cylinder approaches the end of its stroke. For adjustment of the cushion action, a needle valve flow control is built-in.

A cushion consists of a needle valve flow control and a plug attached to the piston. The plug can be on the rod side, in which case it is called a cushion sleeve, or it can be on the cap end side, in which case it is called a cushion spear.

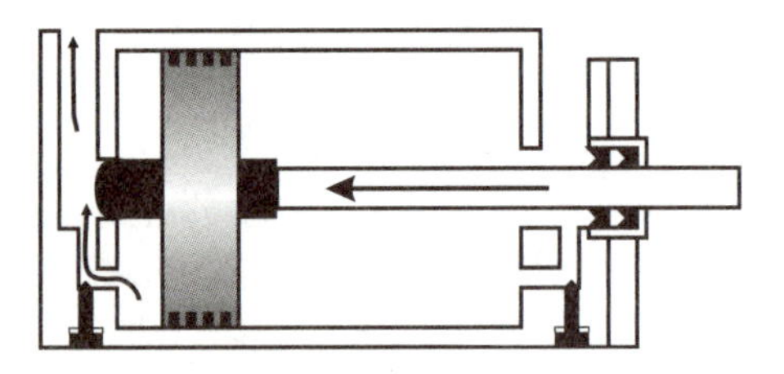

Figure 7-21

Now, when using a cushion to decelerate a moving load, the major consideration from the standpoint of cylinder selection is that the maximum pressure developed by the cushioning device does not exceed the maximum cushion (non-shock) pressure rating of the cylinder.

To determine whether a cylinder is large enough to absorb the energy of the load to be cushioned, the following factors must be known:

1. Total weight to be moved - fixed or variable
2. Maximum piston speed
3. Distance available for deceleration

4. Direction of load horizontal or vertical, thrust or tension
5. Load friction
6. Whether backpressure exists in the cushion cavity before cushioning takes place

Armed with the above information, it can be determined if a standard cylinder with cushions can be selected. However, let us look at some of the above points more closely.

Point 1: The total weight must be determined with the weight of the piston and rod considered. This may be shown in (Fig. 7-4). The total weight is equal to the weight of the piston and non-stroke rod length (Column 1) plus the weight of the rod per inch times the inches of stroke (Column 2) plus the weight to be moved (friction or otherwise).

Bore Dia.	Column 1 Basic Weight (lbs.) For Piston & Non-stroke Rod	Rod Dia.	Column 1 Basic Weight (lbs.) For Piston & Non-stroke Rod
1½	1.5	5/8	.087
2	3.0	1	.223
2½	5.4	1 3/8	.421
3¼	8.3	1¾	.682
4	14.2	2	.89
5	29	2½	1.39
6	41	3	2.0
8	89	3½	2.73
10	115	4	3.56
12	161	5	5.56
14	207	5½	6.73

Total weight = weight of the piston and non-stroke rod length (Column 1) + weight of the rod per inch of stroke x the inches of stroke (Column 2) + the weight to be moved.

Figure 7-22

Example 7-4

A 4" (101 mm) bore cylinder having a 1" (25 mm) diameter rod and a 30" (762 mm) stroke moves a weight with an equivalent load of 100 pounds. What is the total weight?

Solution:

Total weight = piston & non-stroke weight + rod weight + weight to be moved

Total weight = 14.2 + (0.223 x 30) + 100

Total weight = 14.2 + 6.69 + 100

Total weight = 120.89 pounds (537 N)

Point 2: The maximum piston speed may, in some cases, be difficult to determine. The following suggestions are offered to help solve this problem.

1. In horizontal applications where the air pressure is greater than necessary and friction is not a significant factor, it can generally be assumed that the maximum velocity at cushioning will be no more than 20% greater than the average velocity. The exception to this rule is in short stroke cylinders less than 6" in length. In these cases, unless other methods are available for determining maximum velocity, it may be safe to assume that the maximum velocity will be no greater than twice the average velocity.
2. For vertical, rod down applications, the analysis is similar if the speed of the piston at the point of cushioning cannot be mechanically or visually determined. In this case, the maximum velocity at the point of cushioning for all stroke lengths may be no greater than twice the average velocity.

3. The velocities do not consider back pressure, friction and the flow metering occurring in lines. In order to avoid oversizing, these factors should be given ample consideration when analyzing the system requirements. The remaining points are easily understood. Now we must select a cylinder cushion.

Selecting a Cylinder to Cushion a Moving Load

To determine whether a cylinder selected will adequately stop a load without damaging the cylinder or the load being moved, the weight of the load (including the weight of the piston and rod from (Fig. 7-19) and the maximum speed of the piston rod should be used. Once these two factors are known, the "WV" (weight - velocity) graph (Fig. 7-23) may be used).

Enter the graph horizontally at its base with the total weight. Then project vertically to intersect the required speed value from the left. By projecting this intersection parallel to the diagonal lines, a cushion rating number is found.

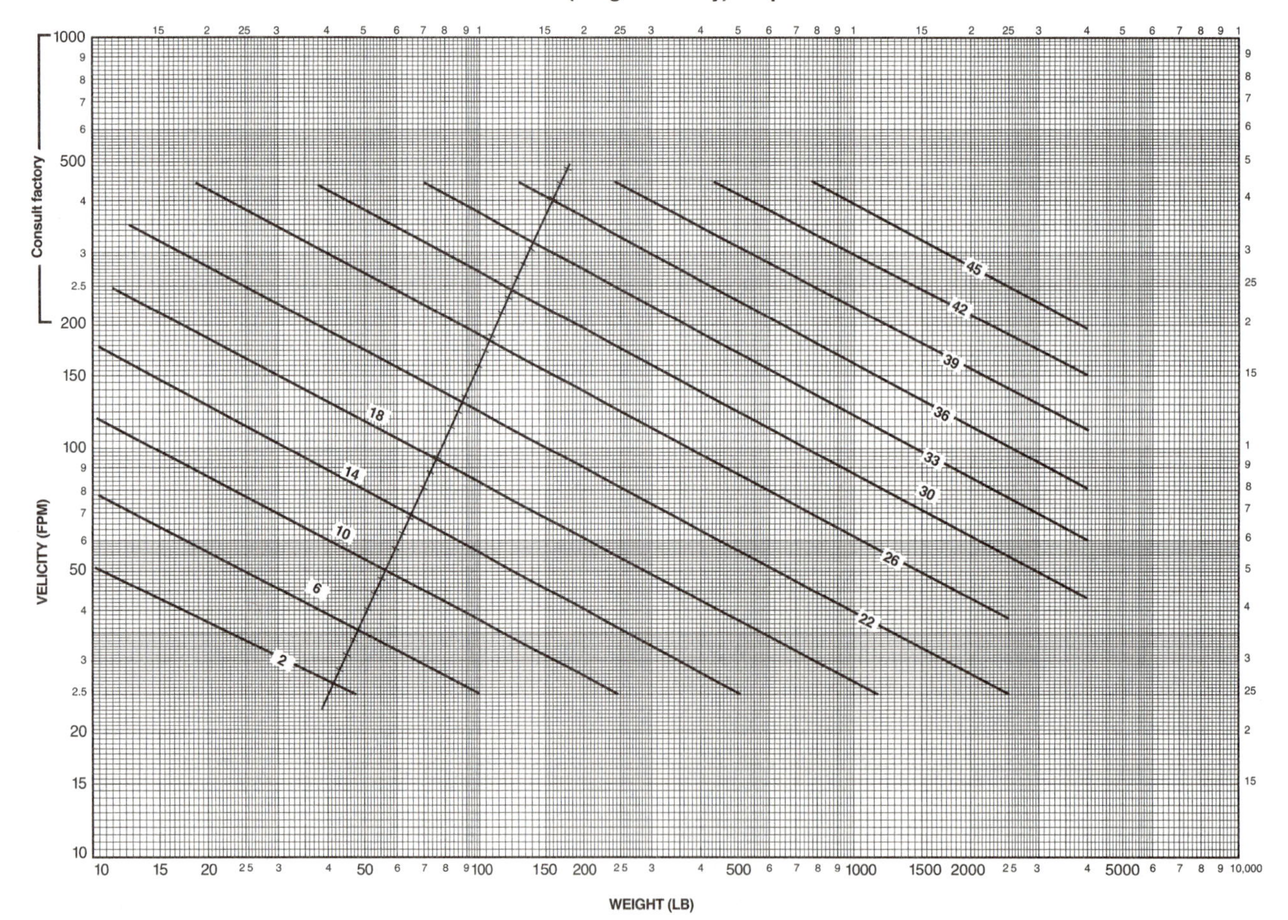

Figure 7-23

At this point Fig. 7-24 is entered and the cushion rating is found for a particular cylinder. If a simple circuit is used, with no meter-out (meter-out being control the air exhausting from the cylinder), use the "No Backpressure Column A" values. If meter-out control is used, use the "Meter-out, Column B" values. As long as the cushion rating found in Fig. 7-24 is greater than the number found in Fig. 7-23, the cylinder will adequately stop the load. If the cushion rating in Fig. 7-24 is smaller than the number found in Fig. 7-23, then a larger bore cylinder should be selected.

Example 7-5

A weight of 85 pounds (228 N) is to be moved at a maximum velocity of 120 fpm (0.056 m/sec) for 25" (635 mm). A 5" bore (129 mm) with a 1" (25 mm) rod has been selected. Will this cylinder's cushion adequately stop the load ?

Solution:

First of all, the total weight must be found. From Fig. 7-22, the piston and non-stroke rod weight of the rod is:

$$\text{Rod weight} = 0.223 \frac{\#}{\text{in}} \times 25 \text{ in.} = 5.7\# \text{ (25N)}$$

$$\text{Total weight} = 29 + 5.7 + 85 = 119.7\# \text{ (523N)}$$

Referring to the "WV" graph (Figure 7-26), the intersection of 120 pounds at 120 fpm is found to need a cushion rating of 22.4.

Referring to Fig. 7-25, Column A, it is noted that the cap end has a rating of 26 and a rod end rating of 23. Therefore, the load and speed can be adequately cushioned by the cylinder selected.

Had the cushion rating determined been larger than 23, the cylinder could have been adequately stopped on retraction but a meter-out would be suggested for the extend stroke. If a meter-out circuit is not practical, a cylinder of a larger size would have to be selected.

NOTE: This exercise dealt with a cylinder in steady state. The conditions at start-up may be different. Therefore initial startup conditions must be checked out before the air is turned on. Now we have a cylinder capable of withstanding the forces imposed on it. We must now examine its speed, which is determined by flow rate.

Series "2A" Air Cylinder Cushion Ratings

Bore Dia.	Rod Dia.	Column A No Backpressure	Column B Meter-out Circuit
1½	Cap End	12	17
	5/8	8	14
	1	3	8
2	Cap End	14	20
	5/8	12	18
	1	9	15
	1 3/8	6	11
2½	Cap End	17	23
	5/8	14	20
	1	14	19
	1 3/8	12	18
	1¾	8	13
3¼	Cap End	21	26
	1	18	24
	1 3/8	17	23
	1¾	16	22
	2	13	19
4	Cap End	23	28
	1	20	27
	1 3/8	20	26
	1¾	19	25
	2	17	23
	2½	17	22
5	Cap End	26	31
	1	23	28
	1 3/8	23	28
	1¾	22	28
	2	20	26
	2½	19	25
6	Cap End	26	31
	1 3/8	26	31
	1 ¾	26	31
	2	24	29
	2½	24	29
	3	22	28
	3½	21	27
	4	20	26
8	Cap End	29	35
	1 3/8	29	35
	1¾	29	34
	2	27	33
	2½	26	32
	3	26	32
	3½	26	32
	4	25	31
	4½	24	30
	5	23	29
	5½	22	28
10	Cap End	33	39
	1¾	32	38
	2	31	37
	2½	31	36
	3	30	36
	3½	30	36
	4	30	36
	4½	29	35
	5	28	34
	5½	27	33
12	Cap End	35	41
	2	33	39
	2½	33	38
	3	33	38
	3½	32	38
	4	32	38
	4½	32	38
	5	31	36
	5½	31	36
14	Cap End	38	43
	2½	37	42
	3	36	42
	3½	36	41
	4	36	41
	4½	36	41
	5	35	40
	5½	34	40

Figure 7-24

Series "2A" Air Cylinder Cushion Ratings

Bore Dia.	Rod Dia.	Column A No Backpressure	Column B Meter-out Circuit
1½	Cap End	12	17
	5/8	8	14
	1	3	8
2	Cap End	14	20
	5/8	12	18
	1	9	15
	1 3/8	6	11
2½	Cap End	17	23
	5/8	14	20
	1	14	19
	1 3/8	12	18
	1¾	8	13
3¼	Cap End	21	26
	1	18	24
	1 3/8	17	23
	1¾	16	22
	2	13	19
4	Cap End	23	28
	1	20	27
	1 3/8	20	26
	1¾	19	25
	2	17	23
	2½	17	22
5	Cap End	26	31
	1	23	28
	1 3/8	23	28
	1¾	22	28
	2	20	26
	2½	19	25

Figure 7-25

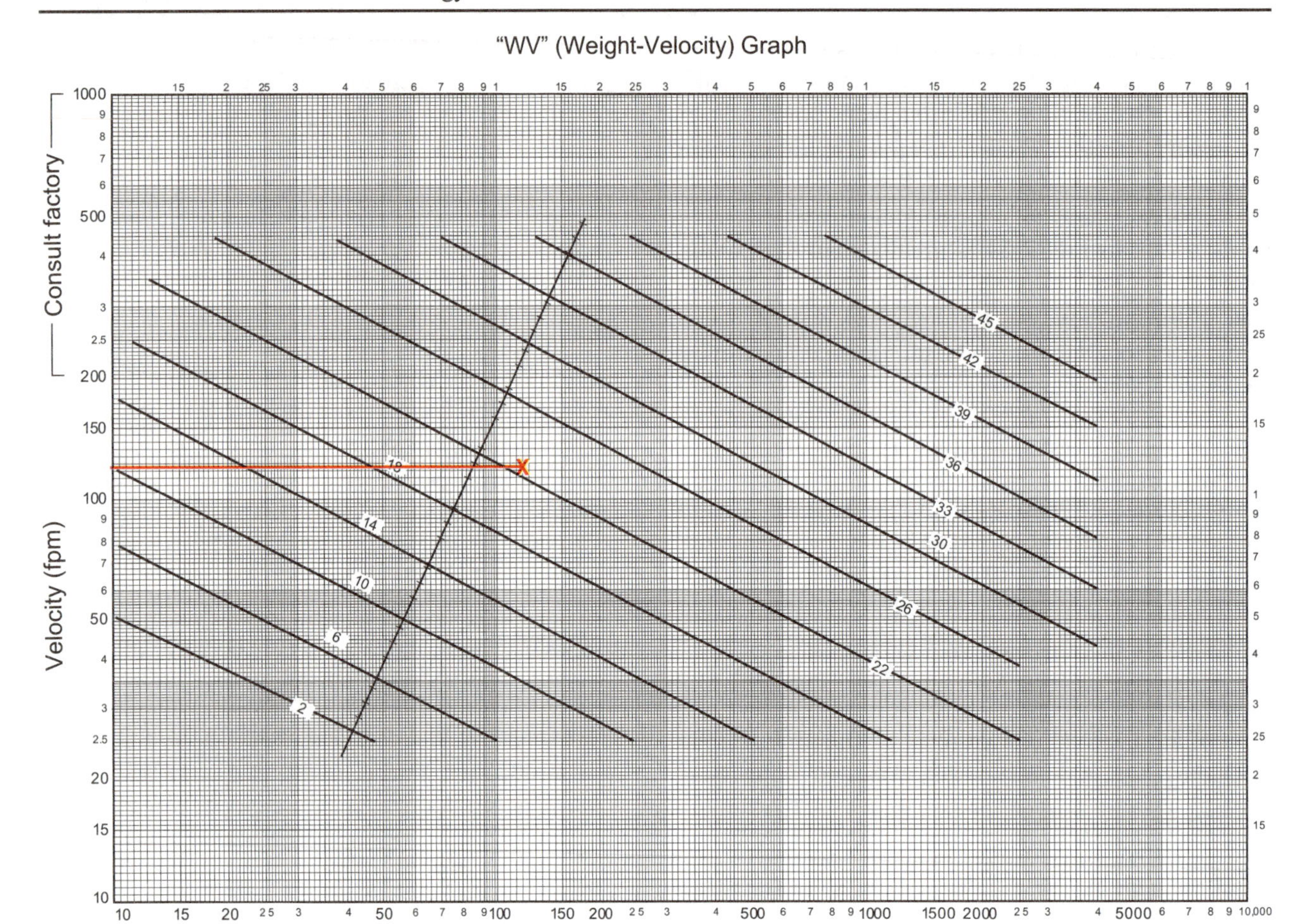

Figure 7-26

Flow Rate Into an Air Cylinder

The flow rate into a cylinder is based upon the amount of air needed to move the piston load and to force out exhaust air from the other side of the cylinder at specified speed.

This may be calculated first by determining the volume of the cylinder.

$$V = A \times S$$

Where:

V is the volume
A is the major or minor area
S is the stroke

For U.S. units:

$$V\ (in^3) = A\ (in^2) \times S\ (in)$$

For SI units:

$$V\ (dm^3) = A\ (cm^2) \times S\ (mm) \times 10^{-4}$$

With the volume calculated, the next step is to find the compression ratio. The compression ratio is the absolute pressure entering the cylinder, divided by the atmospheric pressure.

The equation is: Compression Ratio =

$$\frac{\left[\begin{array}{l}\text{Pressure at cylinder (Psig) +}\\ \text{Absolute Pressure (PSIA)}\end{array}\right]}{\text{Absolute Pressure at Site (psia)}}$$

For S.I units: Compression Ratio =

$$\frac{\left[\begin{array}{l}\text{Pressure at cylinder (Psig) +}\\ \text{Absolute Pressure (PSIA)}\end{array}\right]}{\text{Absolute Pressure at Site (psia)}}$$

We are now ready to calculate the flow rate. Flow rate in pneumatic systems is usually measured in cubic feet per minute (cfm). We know that a cubic foot of air can be at various pressures: 90 psi, 100 psi or 60 psi. But the cubic foot of air referred to in the term cfm generally indicates the air is compressed; it will have a smaller volume depending upon how high the pressure is. As this reduced volume passes through system piping, it is still referred to as a cubic foot of air since that is under normal conditions.

As stated earlier, this term cfm may refer to free air in the same location as the compressor. This is air of the ambient or environment near the compressor's inlet. Since atmospheric conditions vary from day to day and from place to place, free air is not an exact term when comparing flow rates between various systems. For this reason, cubic feet of free air is many times converted to cubic feet of standard air. A standard cubic foot of air is defined as air at an absolute pressure with a temperature of 68°F and a relative humidity and absolute temperature do not vary widely from one installation to another, an abbreviated equation may be worked with. This equation reads:

For U.S. units:

$$\text{cfm} = \frac{V(\text{in}^3) \text{ x Compression Ratio}}{\text{Time to Fill Cylinders(s) x 28.8}}$$

For S.I. units:

$$\text{Flow Rate (dm3/s)} = \frac{V(\text{dm}^3) \text{ x Compression Ratio}}{\text{Time to Fill Cylinders(s)}}$$

Let's work out an example.

Example 7-6

A 4" (101 mm) bore cylinder with 42" (1067 mm) stroke extends in 2 seconds. What is the flow rate into the cylinder during extension if it pushes an equivalent load of 628 pounds (2804 N)? (See level application)

Solution:

Step 1: To begin with, the pressure at the cylinder must be calculated. Using Pascal's Law,

$$Fc = (Pc \text{ x } Ap) - (Pr \text{ x } Ae)$$

Assuming Pr = 0 then

$$Fc = Pc \times Ap \text{ or}$$

$$Pc = \frac{Fc}{Ap} = \frac{628}{12.56} = 50 \text{ psi } (3.5 \text{ bar})$$

Step 2: From here we calculate the cylinder volume.

$$V = A \times S = 12.56 \times 42 = 528 \text{ in}^3 \ (8.54 \text{ dm}^3)$$

Step 3: The compression ratio is:

$$\text{Compression Ratio} = \frac{50 + 14.7}{14.7} = 4.4$$

Step 4 : The flow rate may now be calculated.

$$\text{cfm} = \frac{V (\text{in}^3) \text{ Compression Ratio}}{\text{time(s)} \times 28.8}$$

$$\text{cfm} = \frac{528 \times 4.4}{14.7} = 40.3 \ (18.6 \text{ dm}^3/\text{s})$$

Hence the cfm needed to move the cylinder is 40.3. **This will be the value used later to determine the size of filters, regulators or lubricators.**

Note: Due to the slight difference between cfm and scfm (in the average pneumatic system) we shall use those terms interchangeably.

Practical Cylinder Sizing

As mentioned before, no allowance has been made for "stiction" effects, or for inertia effects of loads applied to the cylinder rod. It is usually best to select slightly oversized (i.e., the next larger bore size) rather than a marginally sized cylinder to ensure continuous operation. Thus, sizing may be based upon the theoretical thrust developed times a design factor of 20% to 50%. This design factor is used to allow for friction and other mechanical losses, loss of line pressure and possible leakages.

The linear motion of a cylinder or the rotary motion of an air motor is not limited to these devices. They can be incorporated and utilized in such things as pneumatic tools.

Pneumatic Tools (Percussive and Rotary)

Pneumatic tools first came on the scene about 1850. They were commonly used to perform the work of rock drilling and chipping. Around the turn of the century, pneumatic hammers were being used for riveting because the art of welding had not yet been developed. Other types of energy transmission systems had been tried but they were not as successful as compressed air for percussive tools. During the same period rotary tools were being developed. In a past war (WWII) period, a wide range of air powered tools were manufactured, including all sorts of semi-automatic and fully automatic devices.

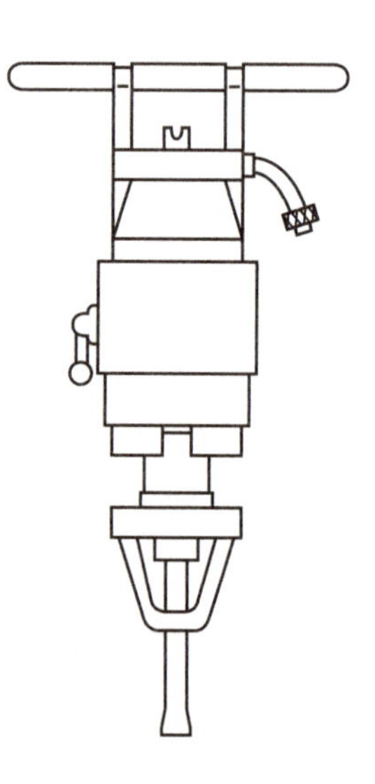

Figure 7-27

Percussive Tools

The percussive tools were among the first to be used with the then more efficient pneumatic transmission system. Tools such as scalers, chipping hammers and rammers all have the basic operating principles derived from the riveting hammer of 100 years ago.

The basic principle of operation is the reciprocating action of a piston with a cylinder. This motion is achieved by progressively loading and unloading a piston. The alternating forces on the piston on a long stroke tool can be obtained with a side valve. In short stroke applications, such as a needle scaler, the piston direction is controlled by piston porting alone.

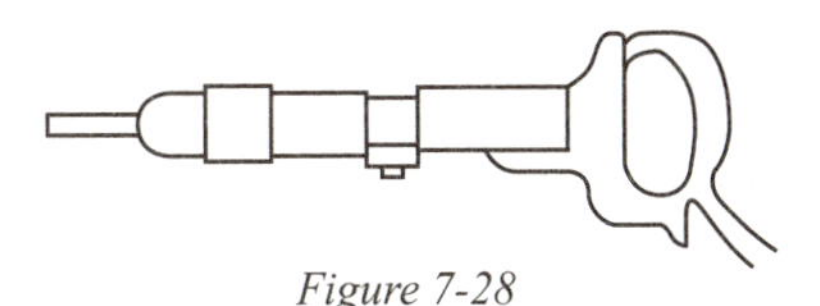

Figure 7-28

Although the principle of operation of all these units is approximately the same, the percussive characteristics can vary greatly.

A particular amount of power can be obtained by varying the mass and velocity of the piston and the number of blows per second. This would seem to tell us that a particular job can be completed by using either a heavy piston with a low number of blows or a light piston with a great number of blows. However, this premise is wrong. Let us look at an example to help explain this.

Consider a riveting machine for ships. The hammer should be capable of delivering a heavy blow to form the head of the rivet while in turn expanding the width. Several choices of hammer exist which will impart the same power. For example, heavy piston - small number of blows (short stroke), or light piston - large number of blows (short stroke). The latter type hammer possesses enough power to shape the head of the rivet satisfactorily. However, because the piston is light, the rapid blows may tend to harden the rivet head beyond permissive tolerances. This will depend upon the type of material makeup of the rivet.

At the other end of the scale are the chipping hammers. When chipping in tough materials, a tool having a long stroke and relatively heavy piston is required. For soft materials, the stroke is reduced which means the impact rate may be increased. For removal of rust scale or weld splatter, a very light piston with corresponding high rate of impact is preferred. The other type of device used in pneumatic tools is the rotary air motor.

Rotary Motors

Pneumatic motors convert the potential energy of compressed air into rotary mechanical energy.

They are compact, relatively light weight units allowing wide ranging variable control of torque and speed. Pneumatic motors are not easily damaged by stalling, reversings or continuous operation. Pneumatic motors operate as a result of an imbalance pressure across a moving or driving member(s) which results in the generation of torque and rotation. Torque indicates a force is present at a distance from a motor shaft. One unit for measuring torque is the lb. in. (N-m). Torque tells us where the equivalent force would be in relation to the motor shaft. The expression which describes torque is:

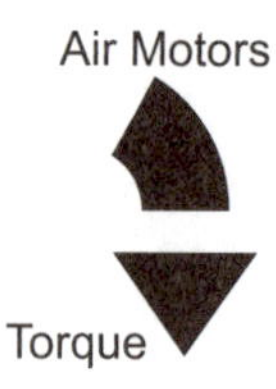

Figure 7-29

For U.S. units:
Torque (lb.in.) = Force (lb) x Perpendicular distance from shaft (in)

For SI units:
Torque (N-m) = Force (N) x Perpendicular distance from shaft (m)

In the illustration, a force of 50 lbs. is positioned on a bar which is attached to a motor shaft. The distance between the shaft and the force is 10 inches. This results in a torque or turning effort at the shaft of 5000 lb.- in. (50 lb x 10 in).

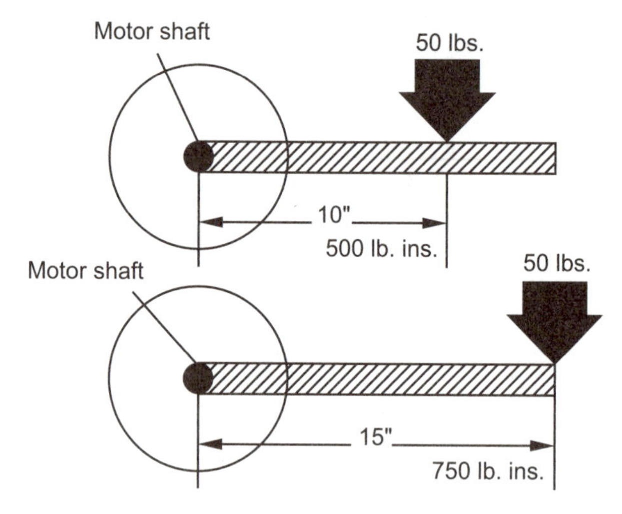

Figure 7-30

If the 50 lbs. were located 15" along the bar, the turning effort generated at the shaft would be equal to a twisting effort of 750 lbs. one inch from the shaft.

From these examples, we can see that the further an effective force is applied from the shaft, the larger the resulting torque at the shaft. It will also be noted that torque does not involve any movement.

Two basic hydrostatic motor types exist: piston and vane.

A hydrokinetic type (turbine motor) is also available. Let us look at the piston type first.

Piston Motor

One of the first low cost efficient air motors was the piston type. Its output ranges between 0.1- 0.2 hp/ in^3 (5 to 10 kW/ dm^3) of cylinder volume, depending upon the design, size, rpm and pressure. These devices are typically low speed, running below 1000 rpm. Since the price-to-power ratio for these devices is high, today you find fewer of this type of motor in the market as compared to the other types such as the vane motor.

Vane Motors

The vane motor develops an output torque at its shaft by allowing compressed air to act on vanes. Vane motors offer significant size advantages as compared to other types. They can produce about 10 times as much power from a given physical size package as the piston types. Typical range of power is 1 to 4 hp/in^3 (50 to 200 kW/dm^3). The rotating group of a vane motor basically consists of vanes, rotor, ring, shaft and a port plate with inlet and exhaust ports.

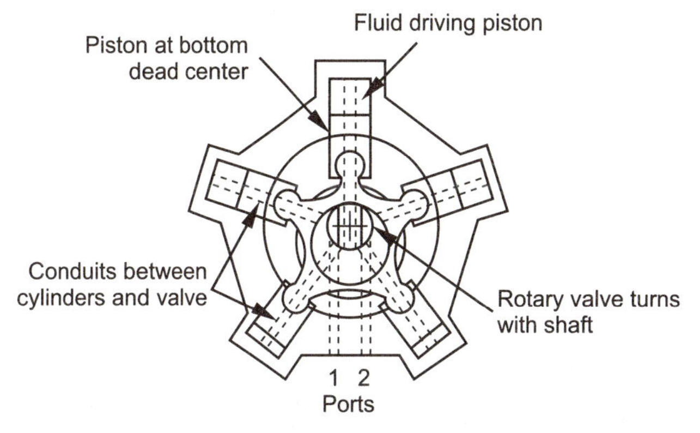

Radial-piston motor
(Courtesy Penton/IPC Publishing)

Figure 7-31

All pneumatic motors operate by causing a pressure imbalance which results in the rotation of a shaft. In a vane motor, this imbalance is caused by the difference in vane areas exposed to air pressure. In our illustrations, with the rotor positioned off center with respect to the ring, the area of the vanes exposed to pressure increases toward the top and decreases at the bottom. When compressed air enters the inlet port, the unequal areas of the vanes result in a torque being developed at the motor shaft.

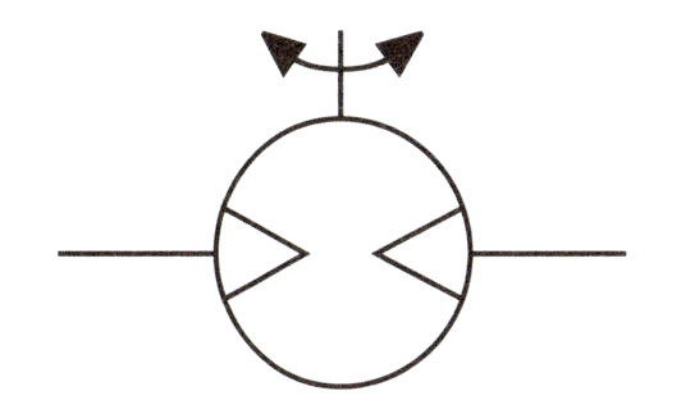

Figure 7-32 Bidirectional motor symbol

It can be seen that the larger the exposed area of the vanes, or the higher the pressure, the more torque will be developed at the shaft. If the torque developed is sufficient to overcome the resisting torque of the load, the rotor and shaft will turn.

But, before a vane motor will operate, its vanes must be extended. Also, a positive seal must exist at the vane tip. In a pneumatic vane motor, vanes are extended by continuously exposing the undersides of the vane to air pressure. This is done by means of passages inside the motor or sometimes light springs are used to insure an initial seal if the driving pressure is built up gradually. There is a third type of air motor. It is the turbine motor.

Turbine Motor

In the turbine motor, compressed air is allowed to pass through a nozzle so that air can expand and therefore its speed can be increased. This fast moving air is then allowed to impinge on a turbine wheel. This motor type has an even higher power-to-weight ratio than the others described. However, turbine motors have very high rotational speeds that are difficult to gear down. Because of this they are limited to special tools such as high speed grinders. The next step is to select the correct air motor for the job.

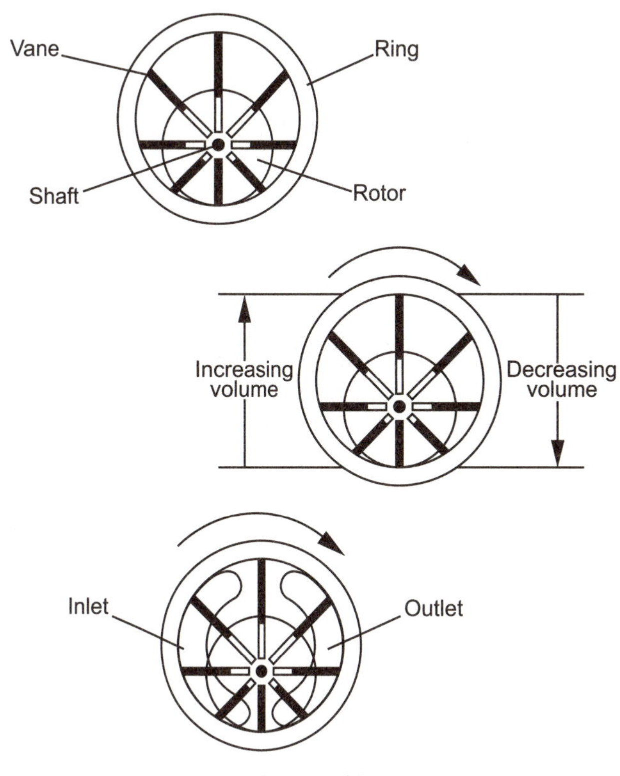

Figure 7-33

Selecting an Air Motor

When selecting an air motor, horsepower, speed (starting and running) and torque are of prime concern. These parameters must be satisfied so that the motor will work in the application. A good rule of thumb for horsepower selection is that the air motor should deliver the required horsepower at about 65% of the available line pressure. In this way full line pressure may be used for starting and overloads. However, many times the type of motor selected for an application is determined by the nature of the application itself. Conventional piston motors, being large and heavy, don't lend themselves to portable tools as lighter vane motors would. On the other hand, hoist and winch applications are easily handled by piston motors. Turbine motors are an excellent choice when very high speeds are necessary.

Example 7-7

A tumbling barrel is used to debur parts. The required speed of the barrel is 5 to 10 rpm. The torque needed to rotate the barrel is 60 ft.-lb. (81 N-m). 100 psig (bar) shop air is available.

What size motor is needed?

Solution:
The torque should be considered first.

For U.S. units:

$$\text{Torque} = \frac{\text{psig x displacement}}{2\,\pi}$$

If we use 65% of the available pressure, our psi value is then 65 psi.

$$\frac{\text{Torque x } 6.28}{\text{psi}} = \text{motor displacement}$$

$$\frac{12 \text{ x } 60 \text{ x } 6.28}{65} = 69.6 \text{ in}^3$$

A motor would be selected that had a displacement of at least 69.6 in^3.

This motor would deliver 60 lb. ft. of torque at 65 psig. If more torque is needed there is an extra 35 psi available to generate the needed torque. The speed would be determined by the flow rate. The power should be calculated next.

$$\text{Power (hp)} = \frac{\text{Torque x rpm x 12}}{63025}$$

$$= \frac{60 \text{ x } 10 \text{ x } 12}{63025}$$

$$= 0.11 \text{ Hp. } (0.085 \text{ kW})$$

The S.I. units:

$$\text{Power (kW)} = \frac{\text{Torque (N-m) x Speed (RAD/S)}}{1000}$$

After the torque and horsepower are known the air motor speed-torque-horsepower curves are entered. A motor is then selected that meets our speed-torque-horsepower needs at our working pressure.

Next we must check the air consumption curve for our motor at our working pressure. This curve will tell us the flow requirements for our motor. Once all of these requirements have been met our motor is ready for work.

Lesson Review

- Air flow (except leakage) passes through a check valve in one direction only.
- A check valve is often used as a bypass valve.
- Cylinders convert pneumatic energy into linear or rotary mechanical energy so that useful work can be done.
- Cushions absorb kinetic energy of the attached moving masses and thus provide protection at the end of a piston's travel.
- A typical stop tube is a collar which keeps piston and rod gland separated at full rod extension.
- The flow rate of air into a cylinder is expressed in terms of cfm (dm^3/s). This is a function of cylinder size, speed and the air pressure.
- Pneumatic tools transform pneumatic energy into linear or rotary extension.
- Percussive tools may be of either medium or long stroke with light, medium or heavy pistons.
- Typical fluid power motors operate by virtue of a pressure imbalance (turbines are a kinetic energy device) across their actuating mechanism.
- Torque is a rotary or turning effort which is expressed in pound-inches (N-m).
- Torque indicates that a force is present at a radical distance from the center of a motor shaft.
- A piston motor is typically large and has a low power-to-displacement ratio.
- The most common type of motor found in industrial pneumatic systems is a vane motor.
- A vane motor develops an output torque at its shaft by allowing the pressure of the compressed air to act on extended vanes.
- Air pressure helps to extend the motor's vanes.
- A turbine motor is hydrokinetic and typically rotates at speeds near 20,000 rpm.

Exercise
Check Valve, Cylinders and Motors
20 Points

Instructions: Match each numbered phrase with one appropriate lettered phrase. There may be more than one correct answer for each match. Phrases may be used more than once.

1.	Converts compressed air into rotary mechanical energy	_____	A. Motor vanes
2.	Separates piston and rod gland bushing	_____	B. Piston motor
3.	Flow in one direction	_____	C. Vane motor
4.	Force at a distance from motor shaft	_____	D. Cylinder
5.	Must be extended by air pressure first	_____	E. Torque
6.	Most common pneumatic motor	_____	F. Check valve
7.	A relatively low speed motor	_____	G. Stop tube
8.	Converts pneumatic energy into linear mechanical energy	_____	H. Cushion
9.	Protects a cylinder from concussion	_____	I. Turbine motor
10.	A hydrokinetic motor	_____	J. Motor
		_____	K. P.O. check

Instructions: Answer the following questions.

11. What pressure will be needed to balance the load? (Piston area = 10 in^2, rod area = 5 in^2)

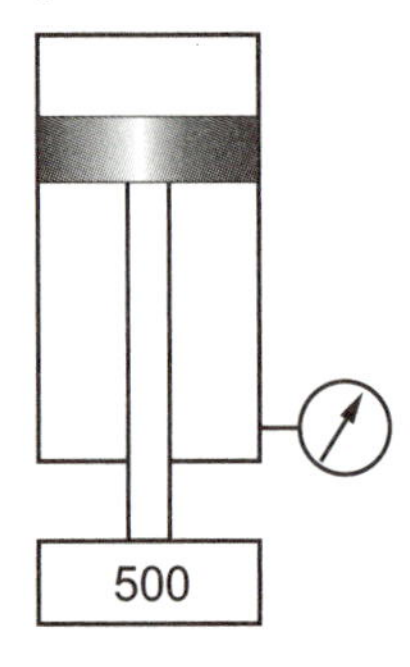

12. A cylinder supported but not rigidly guided, has a stroke of 30". What is the minimum rod size and stop tube length if it pushes a 2,000 pound load?

13. A 4" bore, 2" rod cylinder moves a load of 500 lbs. The average rod speed is 25 ft/min. The cylinder stroke is 5". Will this cylinder be able to cushion the load with no theoretical problem? If not, why not?

14. A 5" bore, 3" rod cylinder moves 24" in 3 seconds. The pressure at the cap and rod end is 50 and 10 psig respectively. What is the air flow rate (cfm) needed to extend and retract the cylinder?

15. If the cylinder in Question 14 was used in Denver (5,000 ft. above sea level), what would be the necessary cfm?

Chapter 8
Directional Control Valves

A directional control valve consists of a body with ports which are connected to internal flow passages by one or more moveable parts which control the direction of air flow. These valves may also function to block or allow air to flow. Other applications in circuits are to control the speed or sequence of an operation. Selection of the higher of two pressures may also be accomplished with some directional control valves.

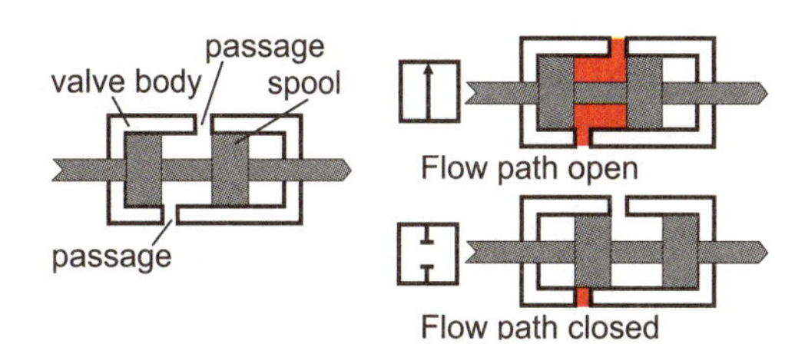

Figure 8-1

Functional Types of Valves

One method of classifying a directional control valve is by the flow paths that are set up in its various operating conditions. Important factors to be considered are the number of individual ports, the number of flow paths the valve is designed for and the internal connection of the ports with the moveable part. Let us begin our directional valve examination with the 2-way valve.

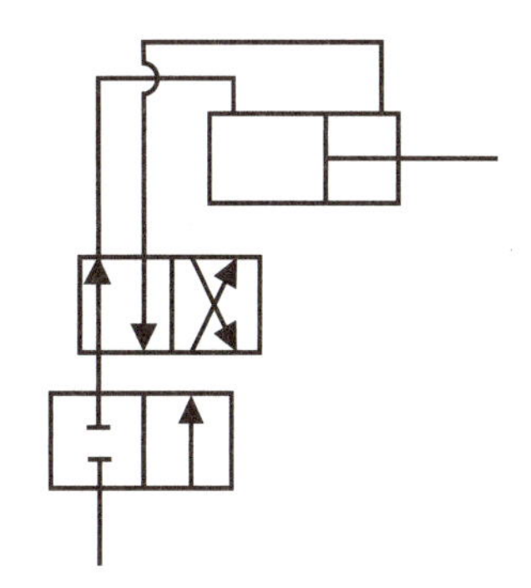

Figure 8-2

2-way Valve

A 2-way directional valve consists of two ports connected to each other with passages which are connected and disconnected. In one extreme spool position, port A is open to port B; the flow path through the valve is open. In the other extreme, the large diameter of the spool closes the path between A and B; the flow path is blocked. A 2-way directional valve gives on-off function. This function is used in many systems to serve as an interlock and to isolate and connect various system parts. The next valve that we will describe is a 3-way valve.

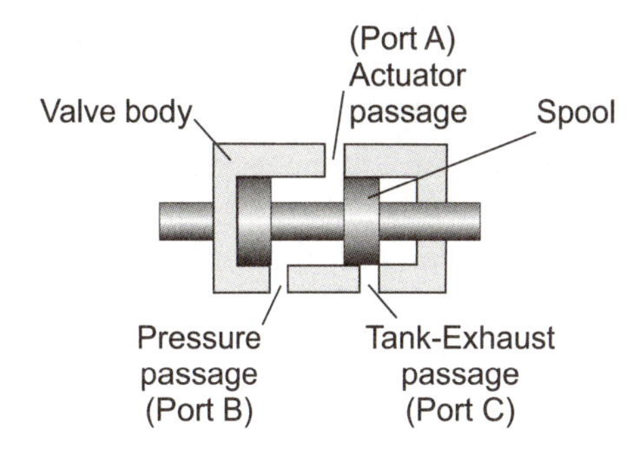

Figure 8-3

3-way Directional Valve

A 3-way directional valve consists of three ports connected through passages within a valve body; they are shown here as port A, port B and port C. If port A is connected to an actuator, port B to a source of pressure and port C is open to exhaust, the valve will control the flow of air to (and exhaust from) port A.

The function of this valve as shown is to pressurize and exhaust one actuator port. When the spool of a 3-way valve is in one extreme position, the pressure passage is connected with the actuator passage. When in the other extreme position, the spool connects the actuator passage with the exhaust passage. The valve can also be used to perform other circuit functions; that is, it can be used to "select" the supply to a device connected to one of two pressures connected to ports B and C.

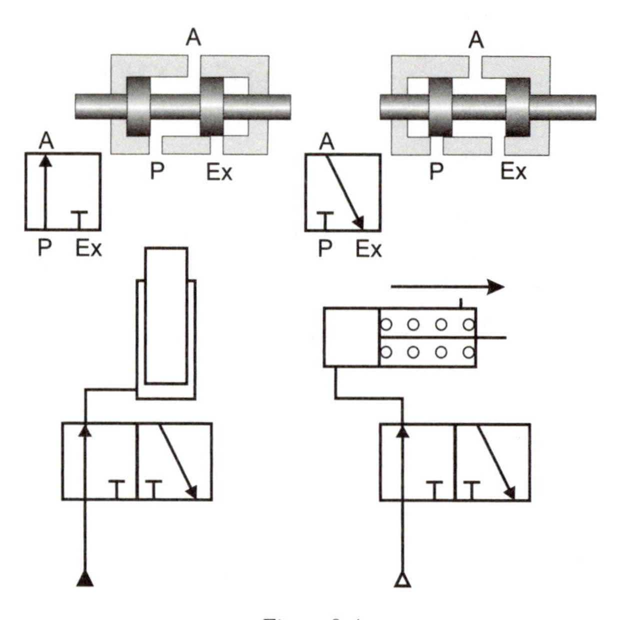

Figure 8-4

3-way Valves in a Circuit

These 3-way valves may be used singly to control single-acting cylinders or in pairs to control double-acting cylinders. In single-acting applications, a 3-way valve directs pressurized air to the cap end side of the cylinder. When the spool is shifted to the other extreme position, flow and pressure to the actuator are blocked. But at the same time, the actuator passage is connected to the exhaust passage. The actuator must be returned by some other method. Also, the 3-way valve may be used in pairs to operate a double-acting cylinder, thus replacing a 4-way valve.

Using paired 3-way valves instead of one 4-way may be for one or more of the following circuit requirements:

1. High cylinder speeds are necessary. This requires air at high inlet pressures and an unobstructed low pressure exhaust. Close coupling of 3-way valves to the cylinder ports cuts down on exhaust back pressure and on pressure drop in lines, allowing greater cylinder velocities.
2. It is desired to save compressed air on high cyclic applications. The use of 3-way valves close coupled to cylinders, eliminates the need to pressurize pipe runs which do not contribute to the work done by the cylinder. This becomes important as larger pipe sizes (½" and up) are used and as higher cycle frequencies are required.

3. Intermediate positions are required. Here two valves are used with their normal position being the pressure port blocked or with both cylinder ports open to exhaust, giving a "float" type cylinder control, when neither of the 3-way valves is actuated. If only the valve at the cap end is actuated, the cylinder will retract. If both valves are actuated, the cylinder will have pressures applied to both ends of the piston. Depending upon the pressure level supplied by each of the valves and the load characteristics, the piston may coast to a stop in some mid-stroke position or move to one end of its stroke.

4-way Directional Valve

Perhaps the most common directional valve in simple pneumatic systems consists of a pressure port, two actuator ports and one or more exhaust ports. These valves are known as 4-way valves since they have four distinct flow paths or "ways" within the valve body.

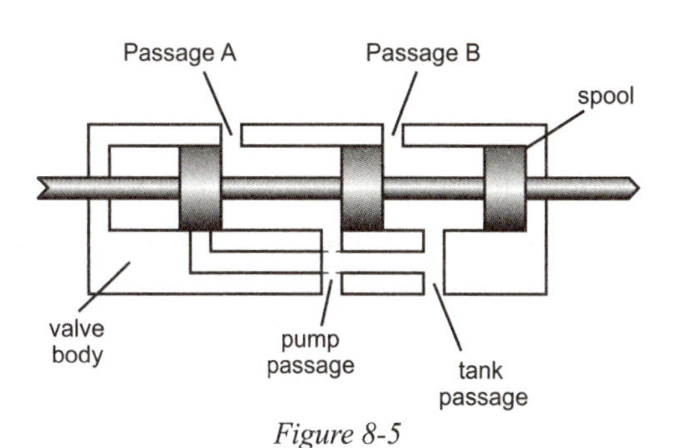

Figure 8-5

A common application of a 4 ported 4-way directional valve is to cause reversible motion of a cylinder or motor. To perform this function, the moveable portion connects the pressure port with one actuator port. This is a 4-ported 4-way valve.

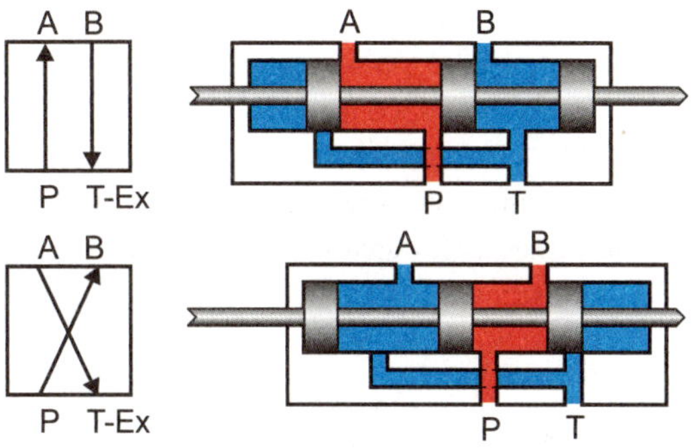

Four-way valves are also available with five external ports. Such valves provide the same basic control of flow paths as the 4-ported version. In the fluid power field this is referred to as a "5-ported, 4-way valve". This type of valve brings all flow paths to individual external ports.

A B
P T-Ex
A B
P T
A B
P T-Ex
A B
P T

Figure 8-6

4-way Valves in a Circuit

The 4-way valve's most common use is to control the motion of double-acting cylinders. Four-way valves may also be used to control a pair of single-acting cylinders or to provide "flip-flop" control of air flow.

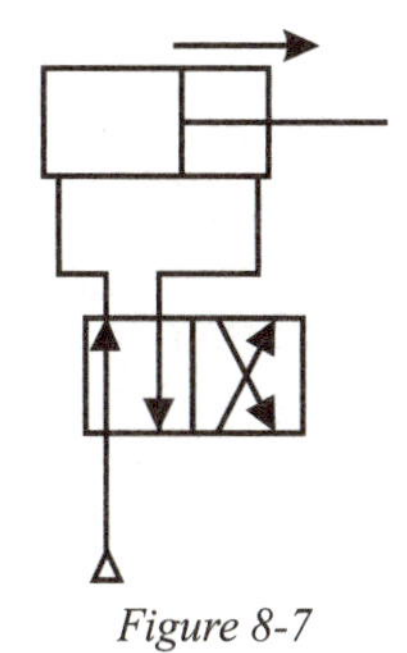

Figure 8-7

However, the 5-ported, 4-way valves may be used to provide dual pressure control. For example, one pressure can be applied to a cylinder when it is extending and a different pressure when its retracting.

Another way to apply a 5-ported, 4-way valve is with a common inlet pressure source and two exhaust ports. This allows individual control of the exhaust from each of the valve's outlet ports. This method is often used to give a type of inexpensive "speed control" of pneumatic cylinders by appropriate restriction of the individual exhaust ports.

Spring Offset

The valves that we have been talking about could be 2- or 3-position type. A 2-position directional valve generally uses an actuator to shift a directional valve spool to an extreme position. Then the spool is returned to its original position by means of a spring. Two-position valves of this nature are known as spring returned or spring offset valves in pneumatic systems.

Normally Open and Normally Closed Valves

Spring returned 2-way valves can be either normally open or normally closed; that is, when the actuator is not energized. In the normally open type, fluid flows may pass through the valve. In a 2-position 3-way valve, since there is always a passage open through the valve, normally closed usually indicates that the pressure passage is blocked when the valve actuator is not energized.

When spring return directional valves are shown symbolically in a circuit, the valve is positioned in the circuit to show its normal at rest condition.

Holding these valves in the actuated position without an external actuating force may cause a control circuit problem. If this is the case, then the use of a detented valve should be considered.

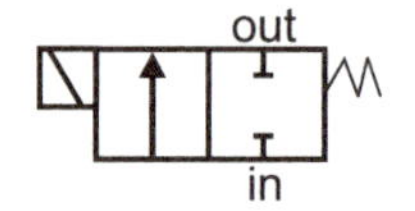

Spring offset - returned, solenoid operated 2-way valve normally closed

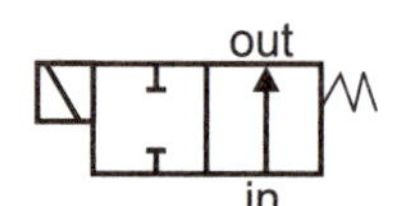

Spring offset - returned, solenoid operated 3-way valve normally open

Spring offset - returned, solenoid operated 3-way valve normally closed

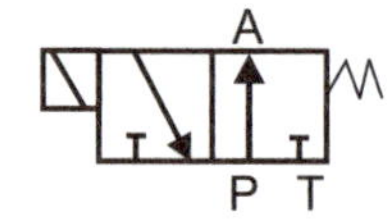

Spring offset - returned, solenoid operated 3-way valve normally open

Figure 8-8

Detents

If either of two momentary actuators is used to shift the spool of a 2-position valve, detents are sometimes used as a holding device to help keep the moveable member(s) in the desired shifted position. However, it may not be sufficient to hold the moveable member against impact, heavy vibration or sudden surge of air.

In our simplified illustration of a detent, the spool is equipped with notches or grooves. Each notch is a receptacle for a spring loaded moveable part. In the detent illustrated, the moveable part is a ball. With the ball in the notch the spool is held in position. When the spool is shifted, the ball is forced out of one notch and into another notch.

Directional valves which are equipped with detents are found more frequently in pneumatic systems than in hydraulic systems.

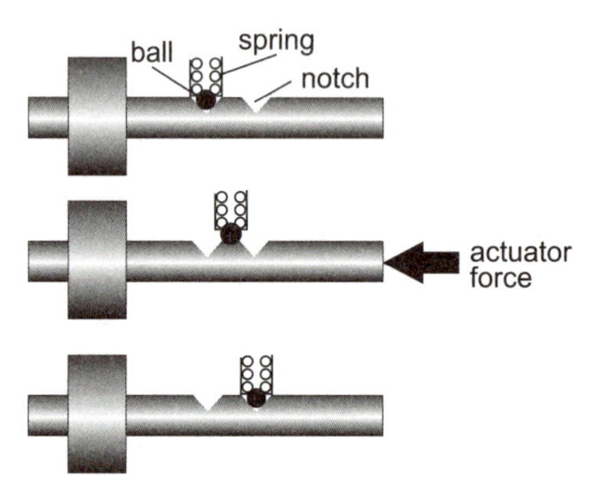

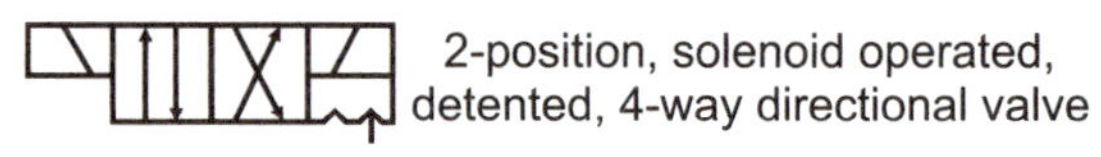

2-position, solenoid operated, detented, 4-way directional valve

Figure 8-9

Three-piston Valves

All the foregoing valve types were 2-position devices providing alternate flow paths - one in the normal position and the other in the actuated position. A 3-position family of pneumatic 4-way values (in both 4- and 5-ported varieties) is commonly available.

Center Condition

In these valves, the two extreme positions of industrial 4-way directional valves are directly related to an actuator's direction. They are the power positions of the valve. They control the movement of an actuator first in one direction and then in the other direction. The center position of a directional valve is designed to satisfy a system requirement. For this reason, a directional valve's center position is commonly referred to as center or "neutral" condition.

Figure 8-10

There are a variety of center conditions available with 4-way directional valves. Some of the more common center conditions are both cylinder ports open to exhaust with pressure (pressure center), both cylinder ports open to exhaust with pressure blocked (exhaust center) all ports blocked (blocked center).

Let us examine each more closely.

Both Cylinder Ports Open to Pressure

First is the pressure center. This type of valve may be used to alternately control the motion of two or three groups of single-acting cylinders. When the valve is in the normal (neutral) position, pressure is admitted to both cylinders, causing extension. If the valve is shifted to either extreme position, one cylinder group will remain extended. When the valve is shifted in the opposite direction, the motion of the cylinders is reversed.

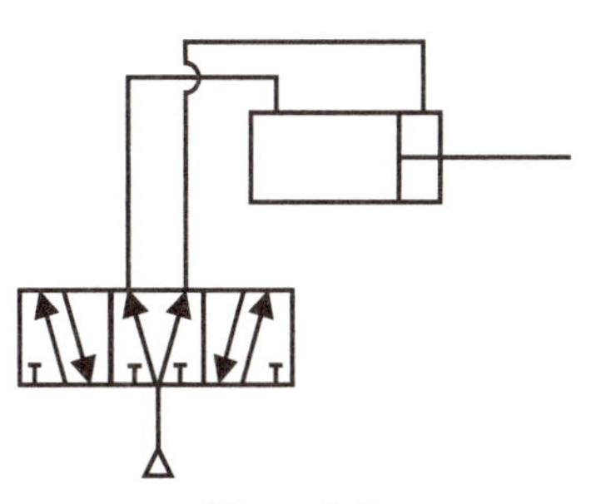

Figure 8-11

Both Cylinder Ports Open to Exhaust, Pressure Blocked

Next we will look at the exhaust center. With this center condition, the cylinder is said to be free to float. In other words, with the valve in its neutral state, both ports of the cylinder are opened to exhaust and the cylinder rod can be moved (subject to internal friction and external loading). When the valve is shifted, the cylinder extends and retracts as discussed before.

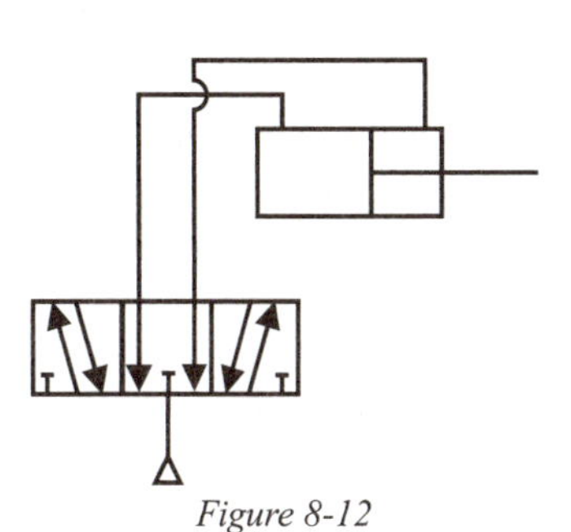

Figure 8-12

All Ports Blocked

Lastly, the blocked center is examined. This center condition blocks all working ports and is often called "closed" or "blocked" center. Depending upon the circuit design and actuator loading conditions, this center condition may provide a holding action on the device to which the valve is attached. With suitable controls and the compressibility of the air taken into account, this type of valve can be used to stop a cylinder along its stroke as it travels in either direction.

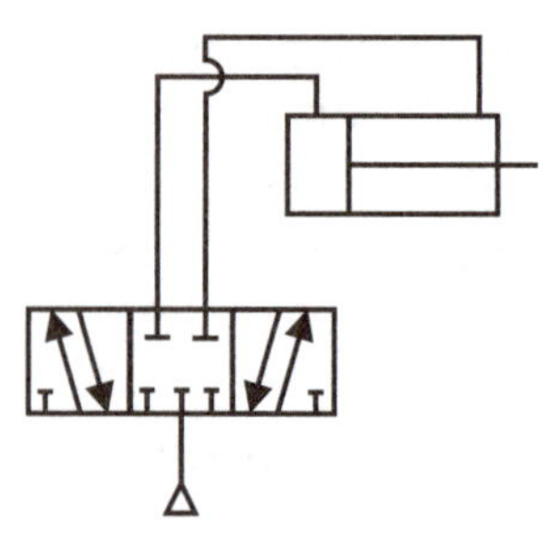

Figure 8-13

However, all of these valves fall into one of two design types.

Basic Valve Design

Basic valve design should be considered when selecting pneumatic directional control valves. Since they control direction, they must be adequately sealed for the application. The sealing function may be accomplished by shear action or poppet-type elements.

Shear Action Valves

The shear action type of valve controls flow by means of an element that slides across the flow path. There are four basic types:

1. sliding plate or rotary disk
2. lapped spool
3. packed spool and
4. packed bore

The first, sliding plate or rotary disc, uses pressure unbalance to force the sealing mechanism (commonly called a D-slide) against a mating surface. The effect is to control the flow of air to and from desired ports and seal the flow from others. These valves can provide 2-, 3- or 4-way action.

The sliding plate principle is available in a modified form called a rotary disc. This utilizes the shear sealing principle of the D-slide valve, but allows control by rotary motion as an actuating means. Correct porting connections are achieved by rotating the D-slide which now may be termed a rotary disc. This can be seen by the illustration at right.

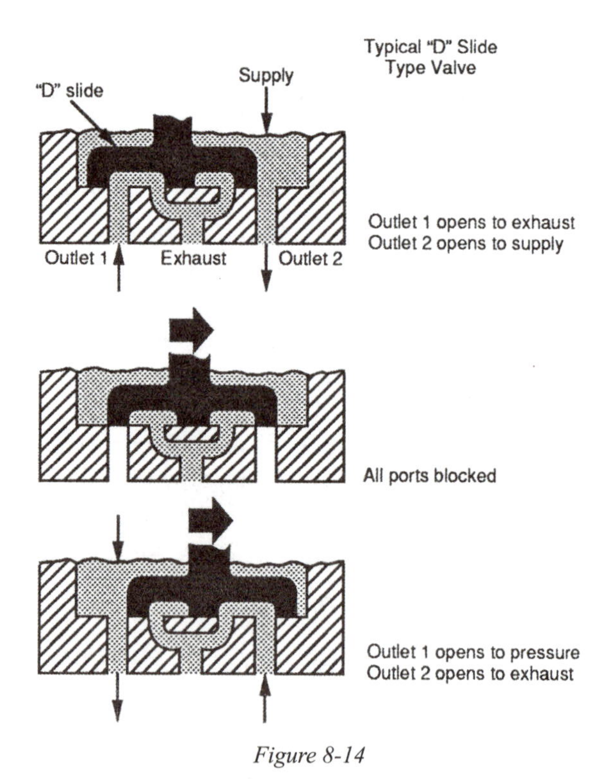

Figure 8-14

Characteristics of Sliding Plate Valves

The following are characteristics of the sliding plate valve:

1. Valve mechanisms "wear in" as they are used. Valves are capable of many millions of trouble-free cycles, even under adverse conditions.
2. Materials may be selected which are not affected by lubricants or contaminants that are present in compressed air lines.
3. Wear tends to keep mating parts in contact, thus controlling leakage over long periods of use.
4. The valve tends to exclude foreign material from lodging between the mating surfaces.
5. The valve is not seriously affected by temperature extremes.
6. The valve can be used as a flow control and directional control.
7. This type of valve has a good air flow throttling in proportion to the movement of the control valve.

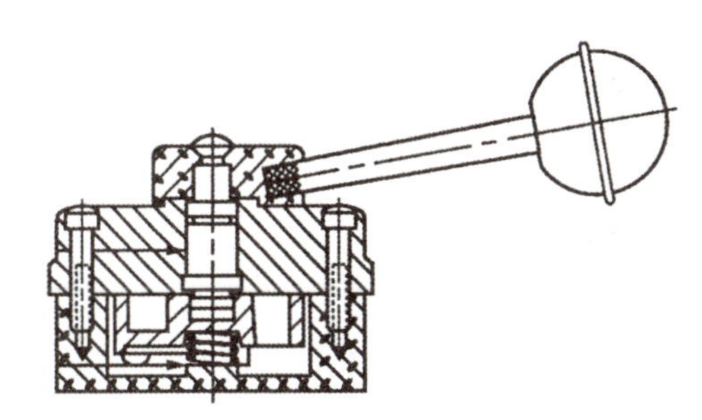

Figure 8-15

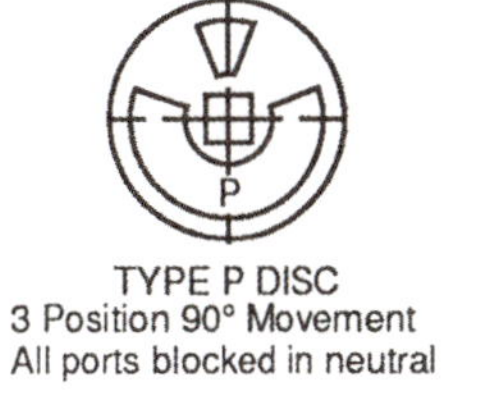

Figure 8-16

Considerations with Sliding Plate Valves

To minimize circuit problems with sliding plate valves, consider these factors:

- Large forces may be required to shift the valve slide, especially at high pressure, when using designs with large pressure sealed areas, which are not at least partially pressure balanced.
- Units may require long actuator stroke lengths to obtain full flow capabilities.
- The valve must be continuously lubricated for maximum life.
- Design may not provide bubble-tight sealing.
- The sealing member (D-slide or disc) can be forced from its mating surface if the pressure under the plate exceeds line pressure. Valves may not be suitable for positioning or stopping cylinders where such pressure conditions can occur.

Packed Spool

Another type of shear action valve is the packed spool valve. This spool type valve is provided with seals to affect almost leak proof sealing. The packed spool valve uses a resilient seal around the spool lands. This spool and seal work in a metal bore.

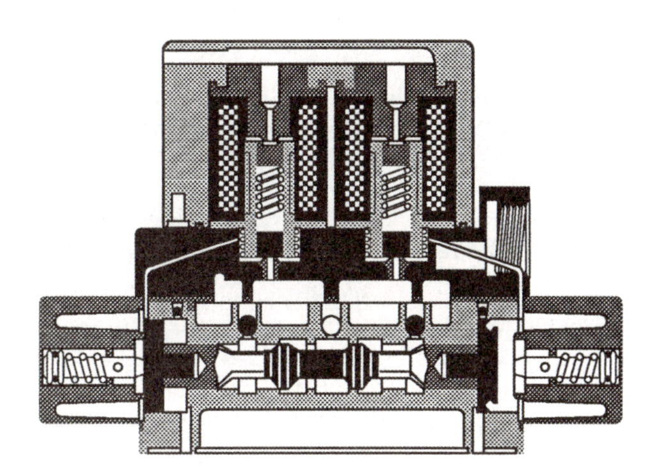

Figure 8-17

Characteristics of Packed Spool Valves

The packed spool valves offer the circuit designer these characteristics:

1. Packed spool design permits maintenance of the seal shape and size.
2. Packed spool valve is less vulnerable to "varnishing" in place than are the lapped spool valves because only small areas in contact or very close fit within the bore.
3. Packed spool valve is less affected by improper installation (improper torquing).
4. Packed spool valve is less sensitive to contamination than the lapped spool valve. The lapped spool valve is covered later in this chapter.
5. Maintenance is less costly from the parts replacement stand point.
6. The valve is of a balanced spool design.

Considerations with Packed Spool Valves

To minimize circuit problems with packed spool valves consider these factors:

- Compatibility of seals and airborne liquids must be checked. Swelling seals require excessive shifting force on the spool. Shrinking seals will increase leakage.
- Some designs use seal materials which tend to adhere to the surface against which they seal.
- Extreme temperatures may cause the seals to change size or become hard.

Packed Bore

There is another very common directional valve. It is the packed bore. The packed bore type valve has a metal spool working in a bore that has several stationary seals which provide isolation between ports.

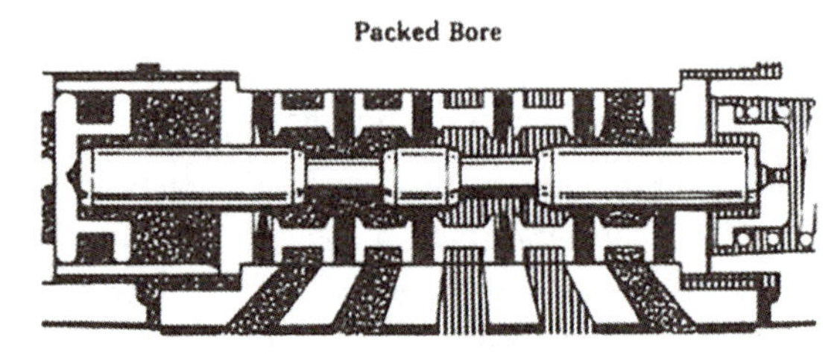

Figure 8-18

Characteristics of Packed Bore Valves

Packed bore valves offer the circuit designer these characteristics:

1. The design is available in a variety of flow path patterns in most porting and actuating configurations.
2. The design results in a balanced spool. Shifting forces are only slightly higher than for metal-to-metal designs. Sudden pressure surges cannot cause the valve to lose its sealing capabilities.

3. The resilient seals make the valve less vulnerable to abrasion by foreign material than metal-to-metal designs.
4. Valve spools have limited areas in contact and less tendency to "varnish" in place than metal-to-metal designs.
5. For greater life, lubrication is suggested.
6. The spool will not bind under ambient temperatures encountered in industrial applications.
7. The seals may be changed without changing mating parts.

Considerations with Packed Bore Type Valves

To minimize circuit problems, consider these factors for packed bore valves:

- Compatibility of seals and airborne liquids must be checked. Swelling seals require excessive shifting force on the spool. Shrinking seals will introduce leakage.
- Some designs use seal materials which tend to adhere to the surface against which they seal.
- Extreme temperatures may cause the seals to change size or became hard.

Packed Spool or Packed Bore

These spool type valves are provided with seals to affect virtually "leak proof" sealing. Packed spool valves are arranged with a spool with resilient seals working in a metal bore. Packed valves have metal spools working in a bore provided with resilient seals.

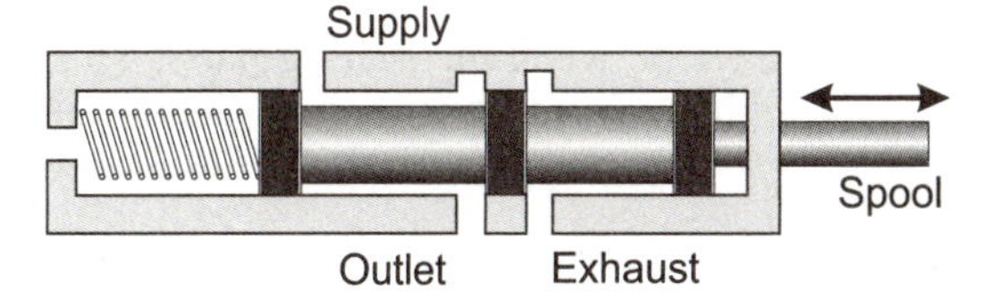

Figure 8-19

Selection of the seals in either case is important. The elastomer must be resilient enough to provide sealing, yet tough enough to provide resistance to abrasion. It must be dimensionally stable under all operating temperatures and in the presence of various airborne fluids.

Lapped Spool Valves

The lapped spool valve is also classified as a shear action type. This design depends upon a close fit between ports to control the flow of air from one port to another. This design does not give "bubble tight" sealing since the spool and bore act like an air bearing.

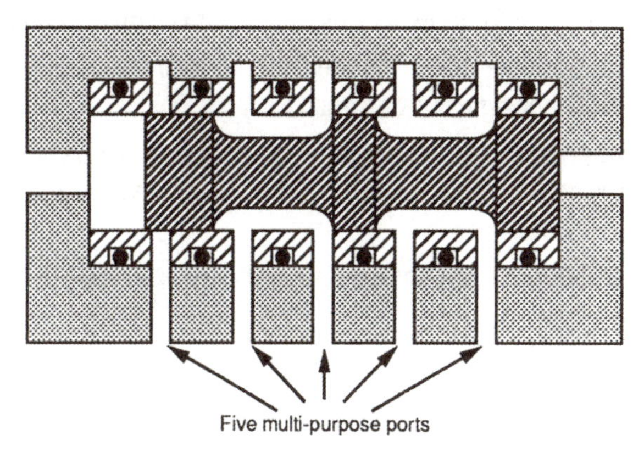

Figure 8-20

The following are characteristics of the lapped spool valve:

1. The design is capable of providing almost any flow path pattern desired, in a variety of porting and actuating configurations.

2. External forces required to shift a balanced spool are low. This is desirable when direct actuation of the spool is required.
3. The force required to position the spool tends to remain constant during a shifting stroke. This is desirable when direct actuation of the spool is required.
4. The design can be made to prevent inter-connection of pressure, outlet and exhaust ports while the valve is being shifted from one flow path condition to another. This ability to eliminate "crossover" flow while this spool is in transit reduces the chance of shift failure when operating at low pressures as a pilot operated valve, and can eliminate spurious signal pulses.
5. Because the spool is balanced, sudden pressure surges resulting from external forces on cylinders cannot cause the valve to lose its sealing capabilities.

Considerations with Lapped Spool Valves

To minimize circuit problems with metal-to-metal spool type valves, consider these factors:

- Long stroke requirements of spools may require excessive travel of mechanical actuators to complete the shifting of the valve.
- Closely fitted parts are vulnerable to the entrance of foreign matter between the mating parts. This can cause rapid wear and leakage, or cause the parts to stick.
- Units of this type require good lubrication and / or filtration. It is sometimes best to run on dry unlubricated air.
- Oxidized airborne lubricant from a compressor or other material carried down an air line may cause the closely fitted parts to "varnish" in place. This occurs particularly when valves are allowed to remain idle for long periods of time, as over weekends.
- Initial cost is high. Maintenance costs are higher than for other types of designs.
- Improper installation (improper torquing) may cause the valve spool to fail to shift due to valve body distortion.
- If the valve is subjected to vibrations, detents may be needed.

Poppet Valves

Up to this point this discussion was limited to shear action valves. We will now discuss poppet types. Directional valves which employ poppets for their moveable part are suited for use where high flows are encountered. Since a poppet opens up a relatively large hole in a short stroke, poppet valves have an inherent characteristic of fast response with minimum wear. Poppet valves for pneumatic service usually employ resilient seals to provide tight sealing. Most poppet valves found in industrial applications are 3-ways. Usually a 4-way poppet valve is made up of two 3-ways.

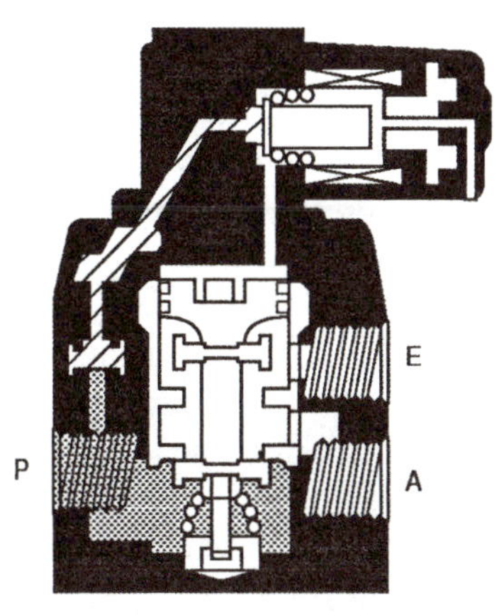

Figure 8-21

Characteristics of a Poppet Valve

Characteristics of a poppet valve would include:

1. Poppet valves offer rapid cycling capabilities. Light moving members require only a short stroke to provide maximum flow opening.
2. The design is reliable and time proven.
3. Short stroke of the valve gives minimum wear and maximum life capabilities.
4. Resilient seals give tight shutoff of the flow path and help absorb the kinetic energy of the moving members.
5. The design is resistant to damage by foreign matter carried through the air line since the seats are self-cleaning.
6. Maintenance of the valve is readily accomplished with inexpensive parts.
7. The valve is not sensitive to air line lubricants or to other materials passing through the valving member.

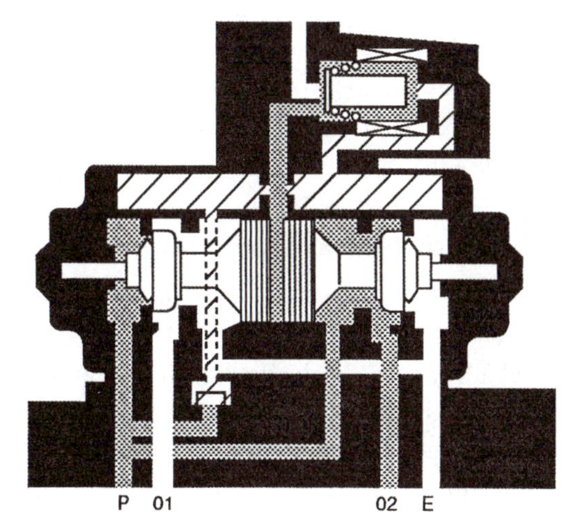

Figure 8-22

Considerations with a Poppet Valve

To minimize circuit problems with poppet type valves, consider these points:

- Poppet designs with directly operated actuators may be bulky in large flow port sizes.
- It is difficult to obtain poppet valves in some flow path configurations.
- The poppet valve design allows flow from the pressure port to escape to atmosphere as the valve shifts. This "crossover" flow may drop the pressure at the inlet port of the valve below the rated minimum operating pressure, causing the valve to malfunction. This shift malfunction has been overcome by building in an accumulator (a miniature air receiver tank) in the pilot air supply.

Now let us look at how these moveable members are positioned.

Directional Valve Actuators

We have seen that a moveable member can be positioned in one extreme position. The member is moved to these positions by mechanical, electrical, pneumatic or manual means. Directional valves whose spools are moved by muscle power are known as manually operated or manually actuated valves. Various types of manual actuators include levers, pushbuttons and pedals.

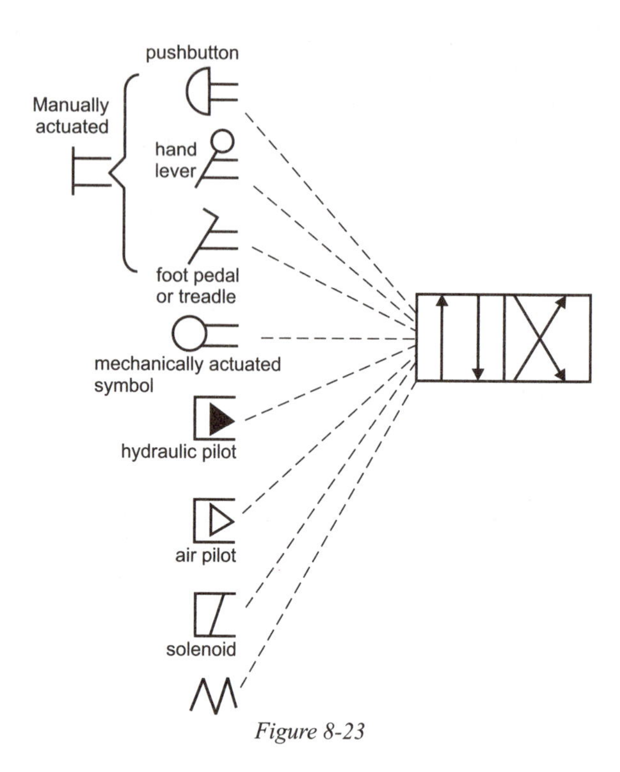

Figure 8-23

A very common type of mechanical actuator is a plunger. Equipped with a roller at its top, the plunger is depressed by a cam which is attached to an actuator. Manual actuators are used on directional valves whose operation must be sequenced and controlled at an operator's discretion. Mechanical actuation is used when the shifting of a directional valve must occur at the time an actuator reaches a specific position.

Directional valves can also be shifted with air. In these valves, pilot pressure is applied to the spool ends or to separate pilot pistons.

Solenoid Operation

One of the common ways of operating a directional valve is with a solenoid. A solenoid is an electrical device which consists basically of a plunger and a wire coil. The coil is wound on a bobbin which is then installed in a magnetic frame. The plunger is free to move inside the coil.

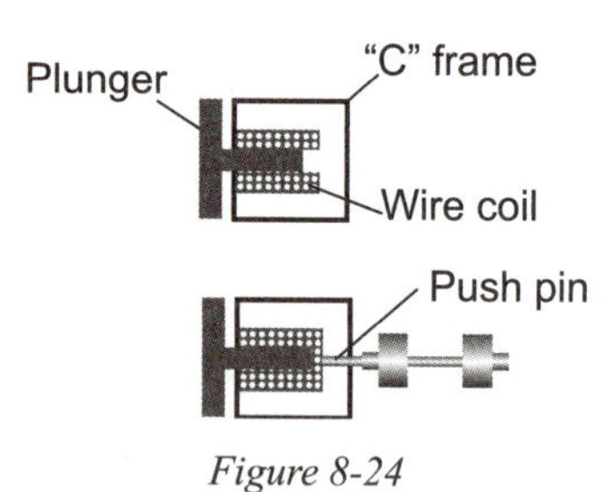

Figure 8-24

When electric current passes through the coil of wire, a magnetic field is generated. This magnetic field attracts the plunger and pulls it into the coil. As the plunger moves in, it can either cause a spool to move or seal off a surface changing the flow condition.

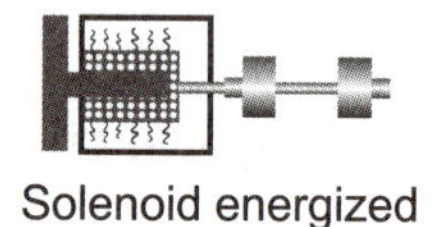

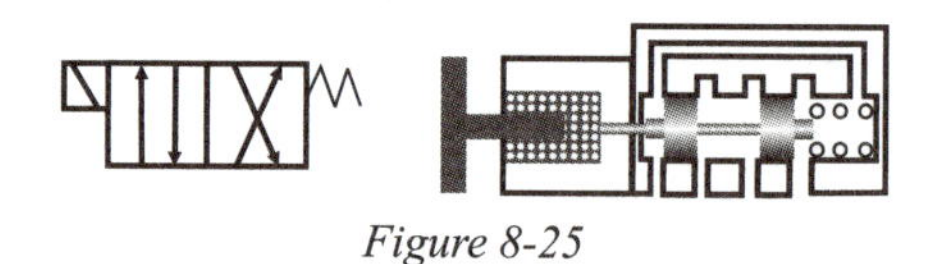

Figure 8-25

Direct Acting Solenoid Valves

When the motion of the solenoid is directly coupled to the shifting mechanism of the valve, it is called a direct operated solenoid valve. Two-position valves with single solenoid actuators usually depend upon a spring or air pressure to return the flow-directing element to its normal condition.

Ex¹ A P B Ex²

Three-position direct solenoid actuated valves rely on solenoid forces to shift the valve member to either extreme position from center. Centering is usually achieved by spring forces or pressure unbalances.

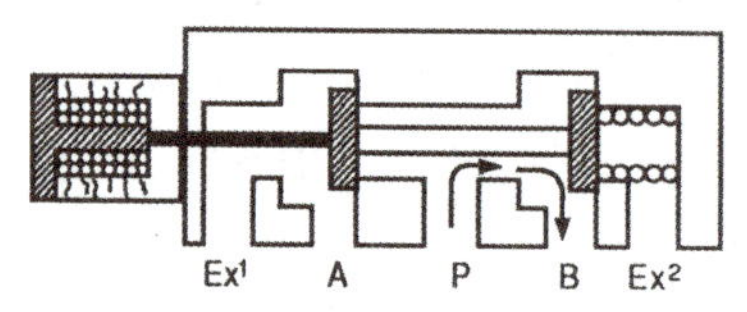

Figure 8-26

Direct operated solenoid valves offer these advantages:

1. The valve will assume the position corresponding to the actuation signal, regardless of the input pressure within the valve or circuit (provided that the maximum is not exceeded). No minimum operating pressure is required.
2. The valve remains in position even if pressure fluctuates to low values.
3. The valve will not tend to malfunction due to restrictions in supply or exhaust.

Pilot Actuated Valves

Another type of valve is the pilot actuator. A pilot actuated valve uses air pressure to move the valve spool. This air pressure may come from a variety of sources.

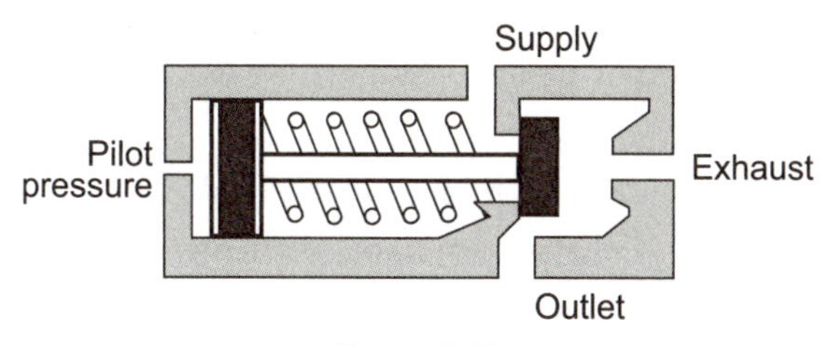

Figure 8-27

Pilot actuated valves are used where actuation is required in remote locations. They are also useful to control low pressures and are a requirement in pressure centered valves. Since actuating forces increase with increasing pressure to the actuating ports, high forces can be generated to shift the valve member. Pilot actuated valves may be internally or externally piloted. A valve is considered to be externally piloted if its air pressure receiver for shifting comes from an external source. If the pilot pressure comes from a source within the valve, it is said to be internally piloted. Let us first examine the internally piloted type of valve.

Internally Piloted Valves

Internally piloted valves use part of the pneumatic energy delivered to the pressure port to position the moveable member within the valve body. Such a valve has a definite minimum and typically a maximum pressure requirement. The application of this type of valve may impose circuit restrictions on the designer. This is due to the need to maintain the minimum specified pressure whenever the valve element requires shifting.

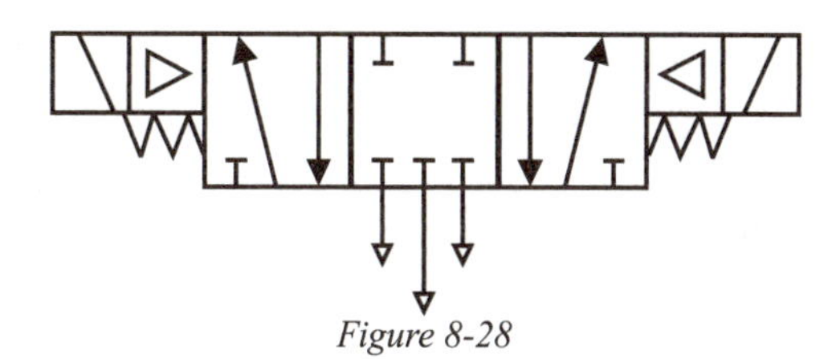
Figure 8-28

The use of properly sized, unrestricted pressure supply lines to the valve will normally avoid problems with valve shifting. Conversely, the use of a supply line which is too small or has built in restrictions may cause the pressure at the valve to fall below the minimum specified operating pressure as the valve is operated. Examples of such restrictions are too small an I.D. in fittings or hose couplings, quick disconnects and sharp bends or kinks in conductors, especially flexible hoses. Problems caused by such conditions may be eliminated by preventing or removing the restrictions or by selecting a valve that is not sensitive to such conditions.

Externally Piloted Valves

Conversely, externally piloted valves require a pressurized source which is derived externally to the valve. This means that a pressure can be selected in such a way that the valve will energize and de-energize in the same amount of time. If slow shifting times are required, a low pressure may be used. If rapid shifts rates are necessary, we can employ high pressures or the rapid exhausting of one of the pilot ports. However, when using high pressure to shift the main valve member, caution should be exercised.

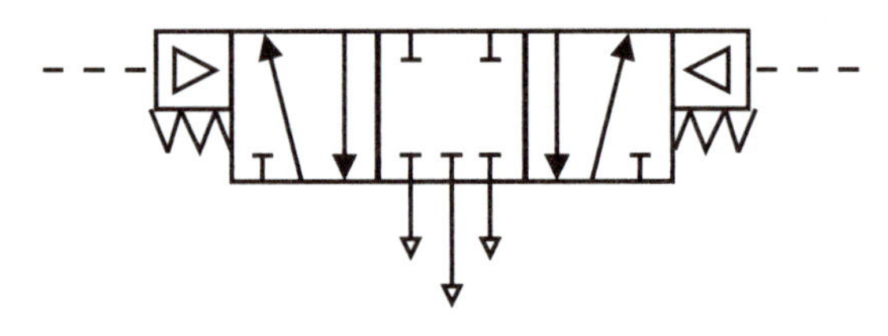

Figure 8-29

Extremely high velocities may cause high impact forces on the valve member, leading to reduced life.

Solenoid-Controlled Pilot-Operated Valves

The forces required to shift large directional control valves may be relatively high, especially if flow forces are unbalanced. To directly operate a valve of this type, large solenoids with high current demands would be necessary. To overcome this problem, many designs make use of solenoid controlled pilot operation. Such designs may provide extremely high flow capability.

Pilot operated valves have these advantages:

1. Alternating current solenoids can operate at low inrush and holding current levels. This results in greatly improved contact life at relays.
2. Low temperature rise at the solenoids decreases the likelihood of solenoid burnout.
3. Small solenoids can be used. This decreases the destructive “pounding” which is often the cause of mechanical failure in solenoids.
4. Solenoids operate small valve elements, which greatly decreases the possibility of failure due to varnish, dirt, etc.

Sub-base Mounting

For convenience these valves can be sub-base mounted. Pneumatic directional valves are frequently fastened to a base to which the system piping and power is connected. This is known as sub-base mounting.

Sub-base mounted directional valves afford ease in servicing, since existing piping does not have to be disturbed for valve installation or removal.

Flow Coefficient – C_v

When selecting these valves, an important consideration is their ability to pass the required volume of air at an acceptable pressure drop. This is referred to as the flow rating. A common method of rating flow is by a C_v "C-sub-v" factor.

The C_v factor is derived from an expression which gives the number of gallons of water per minute that will pass through the valve with a 1 psi differential between the valve's inlet and outlet. It can be calculated with a formula that will be described later.

Typical capacity coefficients for 4-way directional valves are:

Port Pipe Size (NPT)	Flow Rating (C_v)
1/8"	0.2 to 0.8
1/4"	0.5 to 2.0
3/8"	1.5 to 3.5
1/2"	2.5 to 5.0
3/4"	4.0 to 9.0
1"	6.0 to 6.0

In many valve designs, the variation in capacity between different flow paths may vary up to 50%. One manufacturer's ½" port valve may actually pass less flow than another ¼" port valve. Thus, the designer would do well to specify valves that have at least the typical flow coefficients shown.

Sizing a Valve for Flow (For U.S. Units)

There are several C_v formulas in use today. The National Fluid Power Association is currently using the following C_v formula:

$$C_v = \frac{Q}{22.48}\sqrt{\frac{T_1 \times G}{\Delta P \times (P_2 + Pa)}}$$

Where:

C_v = capacity coefficient (a numeral)
Q = flow in standard cubic feet per minute (scfm) @ 14.7 psig, 36% relative humidity
G = specific gravity of the flowing medium (G = 1 for air)
T_1 = absolute temperature (460 + degrees F)
P_1 = inlet pressure (psig)
P_2 = outlet pressure (psig) $P_2 = P_1 - P$
ΔP = pressure drop (psi) static to static pressure
Pa = atmospheric pressure (psia)

NOTES:

1. The effect of relative humidity (RH) on "G" for air is 0.6% over the range of 0 to 100% RH and may be ignored for ordinary test accuracy.
2. This equation is valid for subsonic flow only. To insure subsonic flow (flow velocities below the speed of sound -1100 fps), limit pressure drop so that -

$\frac{P_2 + Pa}{P_1 + Pa}$ is between 0.85 and 1.0

Although this formula is not difficult to work with, we can make use of tables for some of the basic quantities found in the equation. It may be written as:

For U.S. Units

C_v = Flow rate (cfm) x A

Where: Flow rate is the standard cubic feet of free air per minute.

A is a constant (Fig. 8-29), which is a function of inlet pressure and pressure drop.

The flow rate for a particular rotary system is an easy parameter to obtain. This is due to the fact that air tools are typically rated in terms of so many cfm at a particular pressure. For cylinder applications, cfm may be calculated as shown in Chapter 6.

The "A" factor is a variable that is a function of two parameters, inlet pressure (psig) and pressure drop through the interior of the valve. Three "A" constants are given for one primary pressure. Pressure drop is given for three valves. Typically the "A" constant at 5 psi (0.34 bar) drop is used for most applications. On every critical application where maximum efficiency is needed, a 2 psi (0.14 bar) drop should be considered. In many cases, a 10 psi (0.7 bar) pressure drop is not detrimental and can save money and space. Let us try some examples.

Compression Factors and "A" Constants

Inlet Pressure (psig)	Compression Factor	"A" Constants for Various Pressure Drop		
		2 psi ΔP	5 psi ΔP	10 psi ΔP
10	1.6	.155	.102	
20	2.3	.129	.083	.066
30	3.0	.113	.072	.055
40	3.7	.097	.064	.048
50	4.4	.091	.059	.043
60	5.1	.084	.054	.040
70	5.7	.079	.050	.037
80	6.4	.075	.048	.035
90	7.1	.071	.045	.033
100	7.8	.068	.043	.031
110	8.5	.065	.041	.030
120	9.2	.062	.039	.029
130	9.9	.060	.038	.028
140	10.6	.058	.037	.027
150	11.2	.056	.036	.026
160	11.9	.055	.035	.026
170	12.6	.054	.034	.025
180	13.3	.052	.033	.024
190	14.0	.051	.032	.024
200	14.7	.050	.031	.023

Figure 8-30

Example 8-1

An impact tool is being used in a particular circuit. The tool uses 10 cfm (4.72) dm^3/s) at 90 psi (6 bar).

What is the C_v and what type valve is needed?

Solution:

Since the system needed will be used to power an impact tool, a simple 2-way valve is sufficient. The size of the valve is determined by its C_v.

C_v = cfm x "A"

In this application, we have decided to take a 10 psi drop across the valve to save on initial investment. This means that the upstream pressure of the valve is:

Upstream pressure = impact tool pressure + pressure drop

Upstream pressure = 90 + 10 = 100 psi (7 bar)

The "A" factor from the chart for 100 psi inlet is 0.031. The C_v will be :

$$C_v = 10 \times 0.031$$

$$C_v = .31$$

Let us now work a valve sizing problem with a cylinder incorporated into the circuit.

Example 8-2

A cylinder with a 6" (152 mm) bore and 25" (635 mm) stroke is extended at a pressure of 50 psi (3.5 bar). It reaches the end of its stroke in 5 seconds. What size valve should be used?

Solution:

First of all, the cfm must be found.

$$\text{cfm} = \frac{\text{volume (in}^3\text{) x compression ratio}}{\text{stroke time (sec) x 28.8}}$$

The volume of the cylinder may be calculated first.

$$V\ (\text{in}^3) = A\ (\text{in}^2) \times \text{stroke (in)}$$

$$V = 28.26 \times 25$$

$$V = 706.5\ \text{in}^3\ (11.5\ \text{dm}^3)$$

The compression ratio can be found in Figure 8-29 – it is 4.4.

The flow rate becomes

$$\text{cfm} = \frac{706.5 \times 4.4}{5 \times 28.8}$$

$$= 21.6\ \text{cfm}\ (10.2\ \text{dm}^3/\text{s})$$

The C_v can now be calculated. Let us consider that a highly efficient circuit is needed, therefore, a 2 psi drop is suggested. This means the "A" factor is about 0.091.

C_v = cfm x "A"

C_v = 21.6 x 0.091

C_v = 1.96

Sizing a Valve for Flow (For S.I. Units)

There is another way to size a valve for flow. It is by calculating the C_v. The equation for C_v is as follows:

Capacity Coefficient Formulas

$$C_v = \frac{Q}{114.5}\sqrt{\frac{T_1 \times G}{\Delta P \times (P_2 + Pa)}}$$

Where:

C_v = capacity coefficient (a numeral)
Q = flow in standard cubic feet per minute (dm^3/s) @ 7600 mm Hg 20°C, 36% relative humidity
G = specific gravity of the flowing medium (G = 1 for air)
T_1 = absolute temperature (273 + C°)
P_1 = inlet pressure (bar)
P_2 = outlet pressure (bar)
ΔP = pressure drop (bar) static to static pressure
Pa = absolute pressure

Notes:

1. The effect of relative humidity (RH) on "G" for air is 0.6% over the range of 0 to 100% RH and may be ignored for ordinary test accuracy.
2. This equation is valid for subsonic flow only. To insure subsonic flow limit pressure drop so that –

$$\frac{P_2 + Pa}{P_1 + Pa} \text{ is between 0.85 and 1.0}$$

In some applications we must check another valve parameter, that of valve response time.

Valve Response Time

Many manufacturers provide data on the time required to pressurize specific volumes. With 100 psi supply, time required to fill from 0-90 psi and exhaust from 100 psi - 10 psi is measured from instant of energizing, or de-energizing 120v/60 Hz. It stops when a specified percentage of the applied pressure is attained in a load chamber. Some typical performances for relatively fast valves are:

Time (Seconds) to Pressurize or Exhaust Fixed Volumes*

Port Pipe Size (NPT)	12 in³	100 in³	1000 in³
¼"	0.060	0.300	—
3/8"	0.040	0.090	—
½"	0.035	0.075	—
¾"	0.030	0.070	0.55
1"	—	0.065	0.40
1¼"	—	0.060	0.37

*Times are measured from instant of energizing or de-energizing a 115 v. 60 cps AC solenoid. Supply pressure is 100 psig. Volumes are filled from 0 psig to 90 psig, or exhausted from 100 psig to 10 psig.

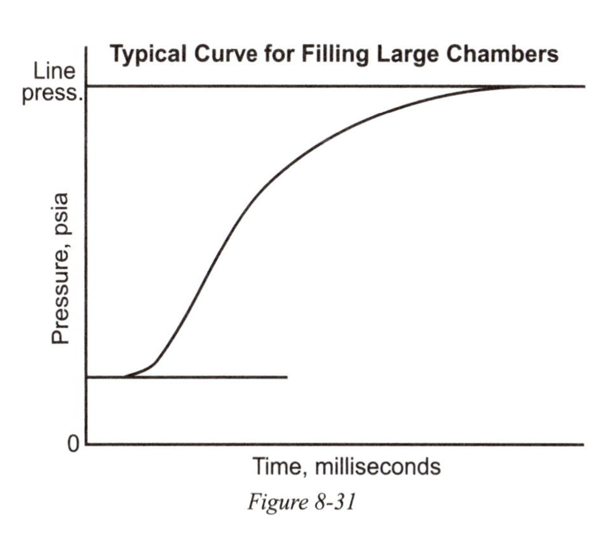

Figure 8-31

The response time of control valves is becoming more important as pneumatic circuitry is used to control rapidly moving machinery or to provide split second accuracy of operation. A valve's operating speed is important in applications where a command or signal given to the valve requires almost instantaneous (and consistent) response of the device controlled by the valve. Generally speaking, the shorter the response time of a valve, the more reliable it will be in its actual application in a high speed or high cyclic circuit. The matching of a valve to a specific response needed should take into consideration the volume being pressurized and the time available for the actuation of the valve itself.

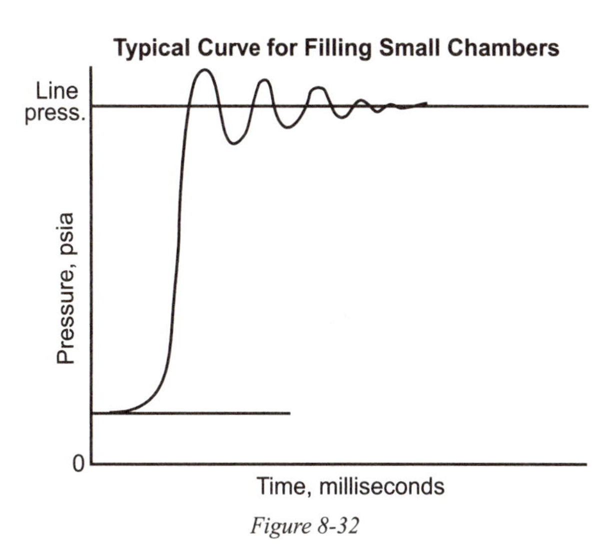

Figure 8-32

Where small volumes are to be pressurized, small ported valves will give faster response than large ported valves. This is because large valves must pressurize sizable internal volumes as well as the loads connected to them. Thus, even though they have large flow capacities, the time required to pressurize them may make large valves appear to have slow response when used on relatively small loads.

Lesson Review

In this lesson dealing with directional control valves we have seen that:

- Directional valves can perform a 4-way, 3-way or 2-way function.
- 4-way directional valves cause a reciprocating motion of a cylinder or motor.
- 3-way valves alternately pressurize and drain one actuator port.
- 2-way valves give an on-off function.
- Directional valves can be operated by mechanical, manual, electrical or pilot pressure means.
- One of the most common ways of operating a directional valve is with an electrical or pilot pressure means.
- 2-position directional valves are many times spring offset or spring returned.
- 2-way and 3-way valves can be normally open or normally closed.
- Directional valves can be equipped with detents which help maintain a shifted position.
- The two extreme positions of a 4-way directional valve are the valve's power positions which are directly related to actuator motion.
- Pneumatic 4-way directional valves are commonly sub-base mounted.
- Pneumatic directional valves are commonly sub-based mounted.
- Larger size pneumatic directional valves are pilot operated.
- There are primarily three center conditions of pneumatic directional valves: blocked center, pressure center, exhaust center.
- Spring centering is a common way of centering pneumatic 3-position directional valves.
- Besides a spool, another very popular moveable part for a pneumatic directional valve is a poppet or disc.
- Size pneumatic directional controls by using the C_v formula.

Exercise
Directional Control Valves
20 points

Instructions: In this exercise, incomplete relationships are set up among directional valve topics which have been covered in the preceding lessons. Complete the suggested relationships by choosing the appropriate word or phrase from the choices at the end of the exercise.

1. Off-on is to 2-way as pressure, exhaust is to ________________________________.
2. Three-position is to spring centered as 2-position is to ____________________________.
3. Actuator movement is to extreme positions as system's needs is to ______________________.
4. One 4-way is to easy piping configuration as high cylinder speeds is to ______________________.
5. Actuator free is to cylinder port's exhaust as actuator held is to ___________________________.
6. Lapped spool is to shear action as poppet is to ______________________________.
7. Solenoid is to electrical operation as hand lever is to __________________________.
8. Low shifting force is to direct operation as high shifting force is to _____________________.
9. Large forces is to large cylinder sizes as large air flow rates in directional control valves is to large ______________.
10. Small valve is to quick response as large valve is to high ___________________________.

A. Capacity

B. Cylinder ports blocked

C. Manual operation

D. Flow rates

E. Two 3-ways

F. 3-way

G. C_v

H. Pilot operation

I. Short stroke - high flow

J. Spring offset

Exercise (cont'd)
30 points

Instructions: Work the following problems

11. A pneumatic grinder needs 20 cfm (9.4 dm^3/s) at 125 psi (9 bar). What size valve (C_v) is needed if initial investment is of prime consideration?

12. A 4" (102 mm) bore, 2" (51 mm) rod extends its stroke of 12" (306 mm) in 0.5 seconds. The equivalent load is 628 pounds (2800 N). The retraction time is unimportant. What size valve (C_v) is needed if efficiency is of greatest concern?

13. A motor runs at 800rpm using 60 cfm (28 dm^3/s) at 50 psi (3.5 bar). Considering a 5 psi (0.35 bar) differential across the valve, what is the necessary C_v?

Chapter 9
Flow Control Valves, Silencers, Quick Exhausts

A flow control valve in a pneumatic system reduces the flow rate in a leg of a circuit and consequently slows actuator speed by acting as a restriction, much as an orifice would do.

Figure 9-1 Adjustable flow control valve symbol

Orifice

An orifice is a relatively small opening in a fluid's flow path. Flow through an orifice is greatly affected by the:

1. size of the orifice (diameter or cross-sectioned area)
2. length of the restriction
3. pressure differential across the orifice
4. temperature of the fluid

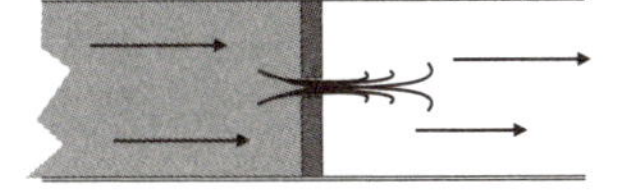

Figure 9-2

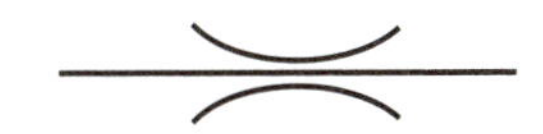

Figure 9-3 Orifice symbol

Orifice Size Affects Flow

The size of an orifice affects the flow rate through the orifice. A common, everyday example of this is a garden hose which has sprung a leak. If the hole in the hose is small, the leak will be in the form of a drip or spray. But if the hole is relatively large, the leak will be a stream. In either case, the hole in the hose is an orifice which restricts the flow of water to the surrounding outside area. The amount of flow which is metered depends upon the size (area and length) of the opening. We shall see that some orifices are fixed in size while others are adjustable.

Fixed Orifice

A fixed orifice is a reduced opening of a nonadjustable size. Common examples of fixed orifices used in pneumatics are a pipe plug or check valve with a hole drilled through the center, or a commercial, factory preset flow control valve.

Adjustable Orifice

In most circuits an adjustable orifice is more useful than a fixed orifice, because it permits adjustments to be made to match the needs of the circuit. Ball valves, globe valves and needle valves are examples of variable orifices. A short description of each is appropriate at this time.

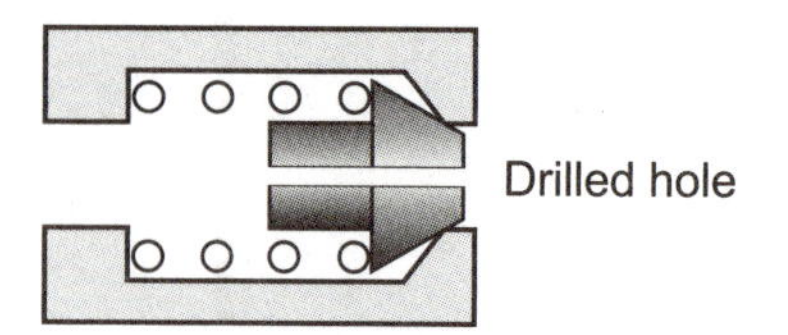

Figure 9-4 Check valve

Ball Valve

First, a ball valve has a flow path which is almost straight through its center. The size of the orifice is changed by turning the handle which positions a ball with a cross drilled hole across the air path.

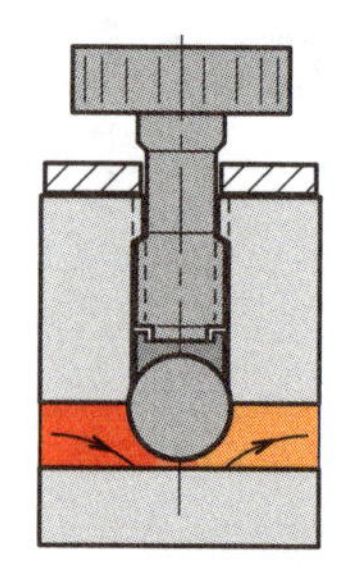

Figure 9-5 Ball valve

Globe Valve

Next, a globe valve does not have a straight through flow path. Generally, the fluid must bend several times as it passes through an opening, which is the seat of a plug or globe. The size of the opening is changed by positioning the globe.

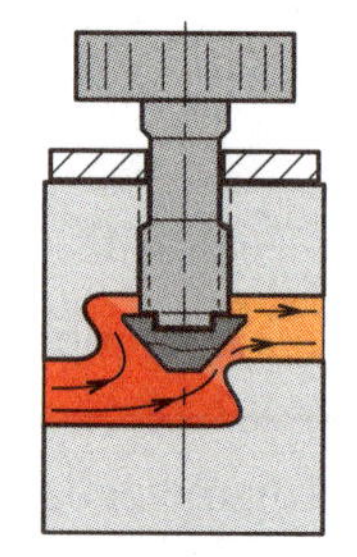

Figure 9-6 Globe valve

Needle Valve

Lastly, the fluid going through a needle valve must also turn several times to pass through an opening, which is the seat for a rod with a cone-shaped tip. The size of the opening is changed by positioning the cone in relation to its seat. The orifice size can be changed very gradually because of the fine threads on the valve stem and the shape of the cone.

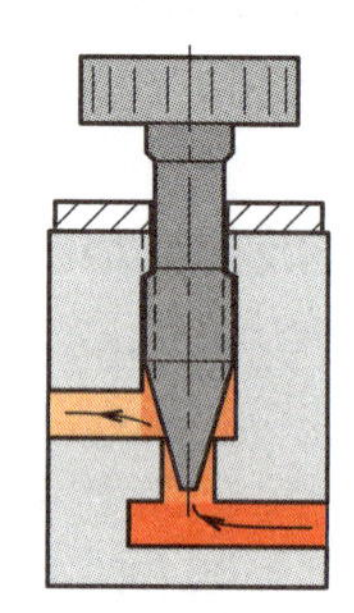

Figure 9-7 Needle valve

Flow Control Valve

The most common type of flow control valve found in a pneumatic system is the needle valve with a bypass check.

Air passing through a needle valve must turn several times to pass through an opening which is the seat for a rod with a cone shaped tip. The size of the opening is changed by positioning the cone in relation to its seat. Orifice size can be changed very gradually because of fine threads on the valve stem and the shape of the cone. The valve may be equipped with an integral check valve which offers little restriction to the reverse or free flow direction. This valve is versatile enough that it can be made into a "sandwich" configuration.

"Sandwich" Flow Controls

This type of device is made to be inserted between a valve and its sub-plate. It adds integral capability to a directional control valve for controlling cylinder speed — SAFETY NOTE: **Due to the compressibility of air, speed control circuits must be examined very closely for all possible failure conditions** — extension and retraction. Individual adjustment of extend and retract speed is obtainable from each needle valve in the sandwich. Improved speed control can be achieved when the valve is mounted very close to the cylinder.

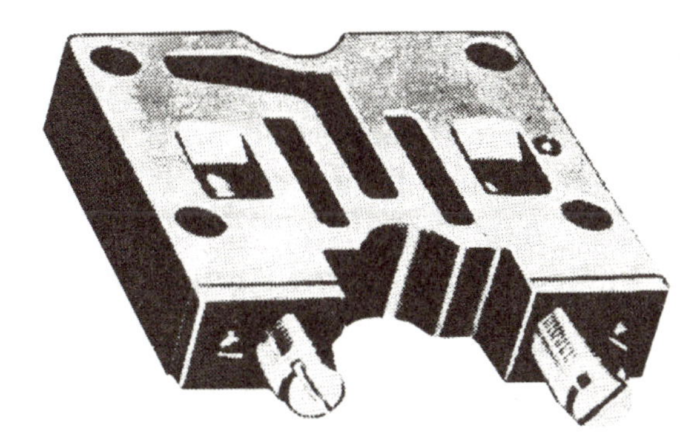

Figure 9-8

Accuracy of a Flow Control

"Sandwich" or inline speed controls in a typical pneumatic circuit are used to restrict or "meter" the air flowing out of a cylinder. This type of usage is termed "meter-out" control. Used in this way, the device provides a degree of control of cylinder velocities in most air cylinder applications, where slow speeds, or close control of speeds under varying load conditions, are not required. SAFETY NOTE: **Due to the compressibility of air, speed control circuits must be examined very closely for all possible failure conditions.** For good "meter-out" speed control, the cylinder load must remain essentially constant. The load includes the force needed to overcome the resisting forces as well as all friction forces. Accurate control of speed in a pneumatic system depends upon many factors. This includes uniformity of the applied load, friction forces, acceleration forces and all other factors which affect the force balance of the cylinder. If the application is such that the cylinder is properly sized and the changes in load and friction forces sufficiently low, then maintaining a fairly uniform backpresssure may be an acceptable method for obtaining cylinder control. The backpressure helps provide a relatively uniform retarding force and helps dampen the effect of "springiness" of the air behind the piston. Therefore, "meter-out" control can be adequate in non-critical applications. SAFETY NOTE: **Due to the compressibility of air, speed control circuits must be examined very closely for all possible failure conditions.**

Figure 9-9

If stable velocity control over the cylinder stroke is required, a more sophisticated circuit tool is needed. Usually a hydraulic dashpot (hydraulic feed device) is used. Properly sized for the application and designed to prevent the formation of wear debris and provided with adequate filtration to prevent silting at the control orifices, this device can accurately control cylinder velocities down to the 1-2"/min. (0.42-0.84 mm/sec.) Our next step is to examine the different flow control placements in the typical pneumatic circuit.

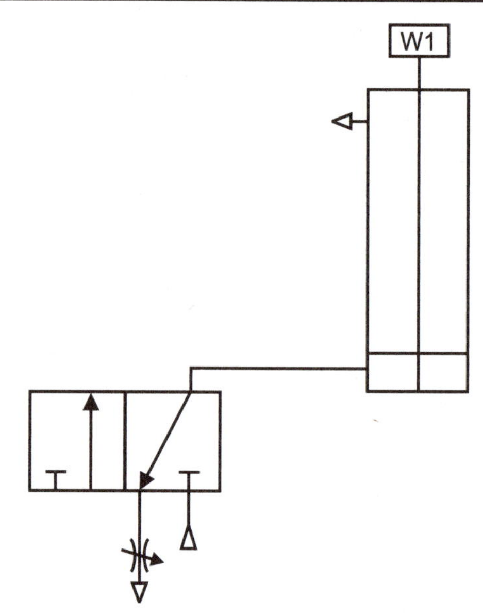

Figure 9-10

Placement of a Flow Control in a Single-Acting Application

Pictured is a single-acting cylinder, flow control and 3-way directional control valve. The cylinder will be able to move at an unrestricted speed in the upward direction. Upon releasing the directional control valve to the "at rest" position, air exhausting from the cylinder must pass through the variable restriction (flow control) and the retraction rate will be controlled. A more common speed control circuit would contain a double-acting cylinder.

Controlling the Speed of a Double-Acting Cylinder Employing a 4-ported, 4-way Valve

Shown are two circuits controlling the rod speed of a double-acting cylinder in one direction only. Both systems use a 4-ported 4-way valve for directional control. In Figure 9-11extension speed is controlled; in Figure 9-12, retraction speed is controlled. In both cases the ratio of flow into the cylinder is controlled by how fast the air is allowed to exhaust through the flow control.

If independent control in both directions is needed, two controls must be used. Here flow contol (1) (Figure 9-13) controls the speed of extension while flow control (2) controls the speed of retraction. Note that flow controls are placed between the cylinder and directional control valve, preferably as close as possible to the cylinder. Greater versatility can be achieved through the use of a 5-ported, 4-way valve.

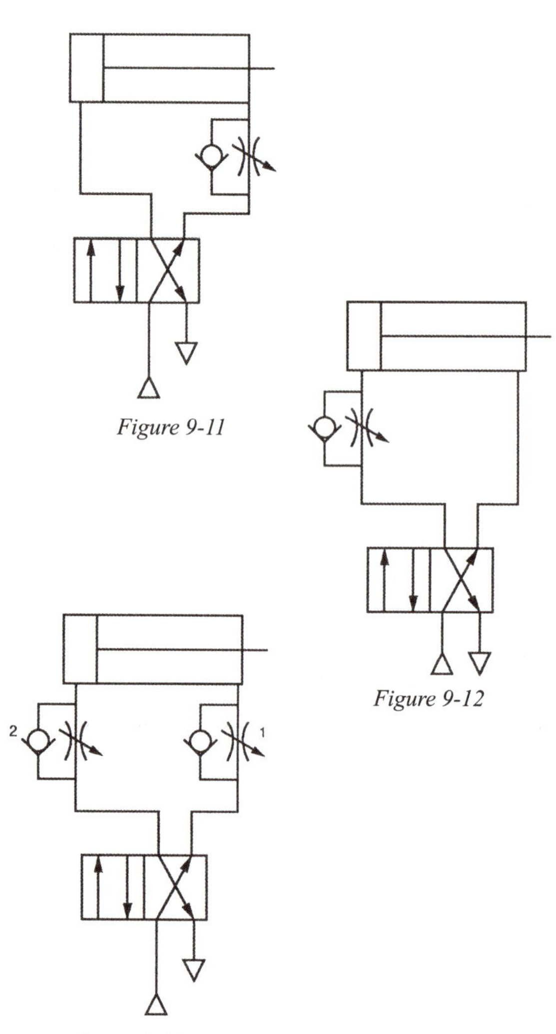

Figure 9-11

Figure 9-12

Figure 9-13

Controlling Speed of a Double-Acting Cylinder Employing a 5-ported, 4-way Valve

As discussed earlier, 5-ported 4-way valves are also available. This valve provides two independent exhaust ports. In Figure 9-14 independent flow controls may be mounted in each exhaust port to control cylinder exhaust and thus speed in both directions.

When the valve is energized (Figure 9-15), the supply is directed to cylinder port (A) and the exhausting air is passed through a needle valve (1).

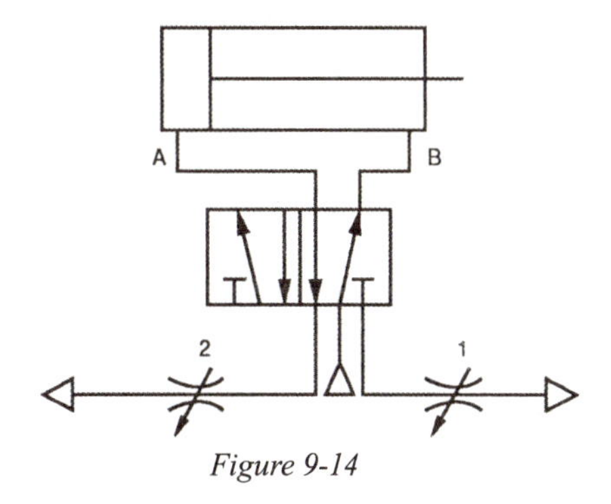

Figure 9-14

During retraction (Figure 9-16) supply is directed to cylinder port (B) and needle valve (2) controls the exhausting air. This directional control valve has a definite advantage. First, it eliminates the need of a bypass check valve being incorporated into the body of the flow control valve. Second, this circuit also offers the opportunity of fewer connections since the needle valve can be fastened into the valve exhaust ports and not in the connecting lines. On the other hand, the control of the cylinder is not as good as when the control orifice is closely coupled to the cylinder. This is because of the need to bring the air up to the "controlled pressure" before control can be achieved. The long lines (volume of line), as compared to the cylinder control volume, will greatly affect when the control effect takes place. Also, it may be more difficult to wring the oil out of the exhaust air.

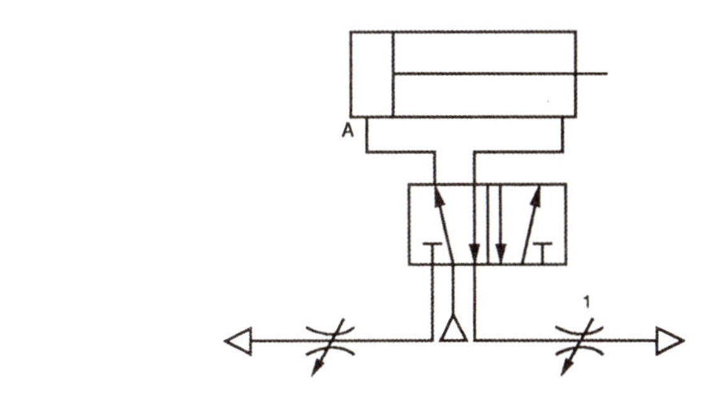

Figure 9-15

Multiple Speed Controls

Many times two speeds are required as a cylinder moves through its stroke. For instance, let us say that we have a cylinder with a stroke of 20 in (508 mm). A fast forward speed of 20 ft/min. (0.10 m/s) is needed for the first 10 inches (254 mm) with a speed of 5 ft/min (0.025 m/s) for the remainder of the stroke. The desired return speed is 25 ft./min (0.125 m/s). It is evident that we need three speed controls and at least one directional control valve. By using a 5-ported, 4-way, the fast forward and return speeds can be obtained by installing needle valves in the two independent exhaust ports. This is shown in Figure 9-16. The slow speed forward remainder must now be obtained. For this section a valve capable of sensing the cylinder rod position must be added. A normally-open 2-way cam operated valve can be used. It is placed in parallel with the slow speed flow control so that as the first 10" (254 mm) of forward stroke is completed, flow is directed through the newly placed flow control. It may be a difficult task to place the cam in the correct position. This is because the valve must be actuated soon enough to allow for the build up of the backpressure required for the final traverse velocity. Let's go through the circuit one step at a time.

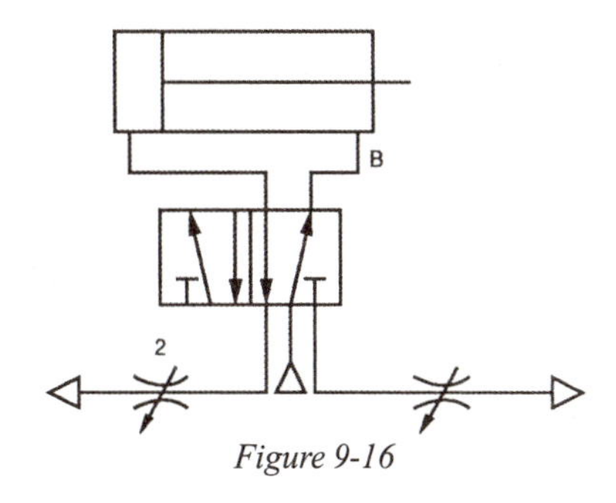

Figure 9-16

The de-energized circuit is shown in Figure 9-17.

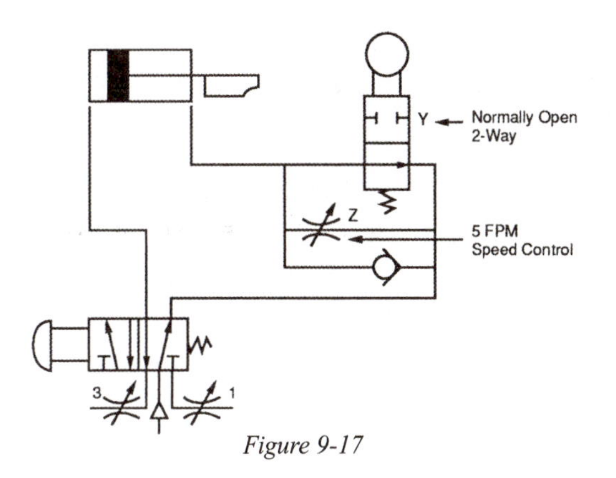

Figure 9-17

When the main valve is energized, air is directed to the "A" port of the cylinder (Figure 9-18). Air exhausting from the "B" port flows through both cam valve "Y" and needle valve "Z", (which is independently set to provide a motion of about 5 ft min (0.025 m/s). Most of the air travels through the open cam valve and then passes through needle valve "1", which is adjusted after "Z" has been set, so that the two valves exhausting together provide a motion of 20 ft/min (0.10 m/s).

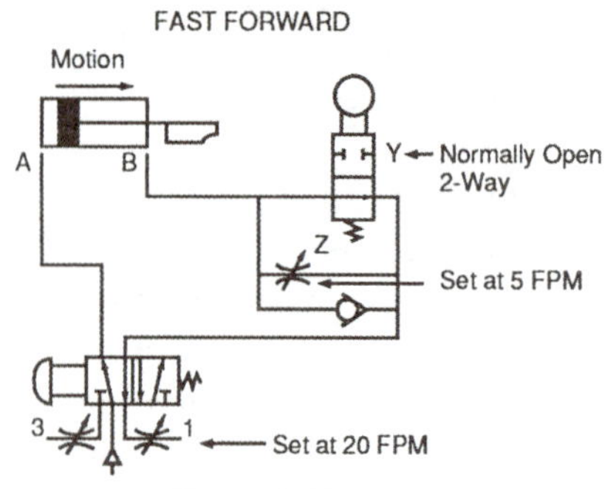

Figure 9-18

When 10 in (254 mm) of forward stroke is sensed, the cam valve shifts to its closed position (Figure 9-19). This causes all the air exhausting from the cylinder "B" port to pass through needle valve "Z". Since the valve was independently set to provide motion of about 5 ft/min (0.025 m/s), the speed of the cylinder is gradually reduced until the new pressure balance is established.

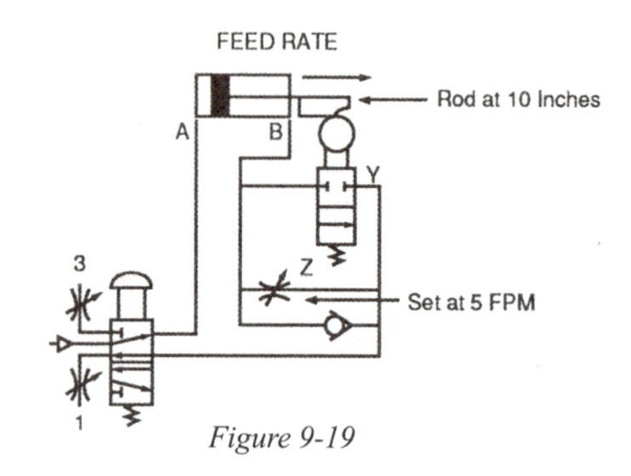

Figure 9-19

Remember, the exact point of the stroke where this occurs is a variable, depending on the volume in the exhaust lines between the cylinder "B" port, the line to the cam valve and the line to the "Z" needle valve. The entire volume in these lines must be brought to the new equilibrium pressure before the reduced speed can be realized. For retraction (Figure 9-20), the directional valve is released, thus directing supply air to the cam valve "Y" which is closed, needle valve "Z" and the check valve. Since the check has the least resistance, a nearly unrestricted flow is fed into cylinder port "B". Air exhausting from "A" is passed through the directional control valve and into needle valve (3) which is set at 25 ft/min (0.125 m/s), the required retraction speed.

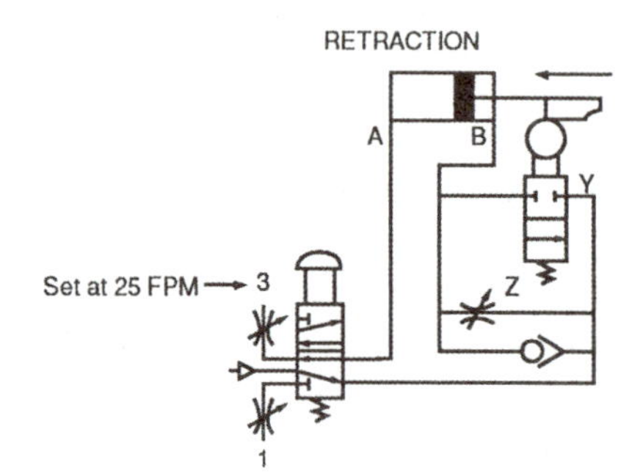

Figure 9-20

Once again, since needle valve (1) is not closely coupled to the cylinder, some problems of control may exist.

An improvement is achieved by placing needle valve "1" close to the cylinder (Figure 9-21). Now, even though adjustments are still interactive, problems with control may be diminished because trapped volumes have been reduced.

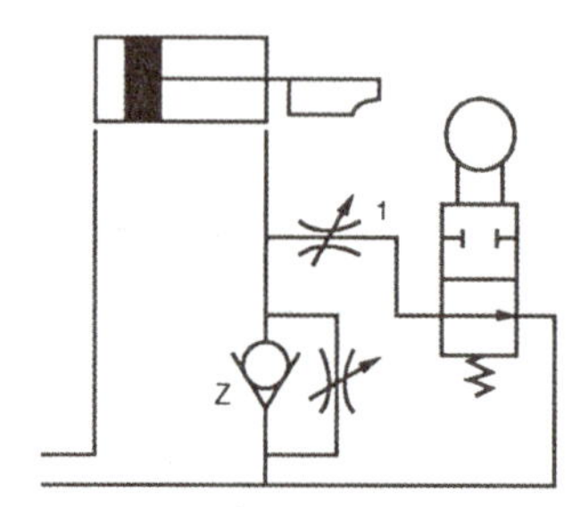

Figure 9-21

Another change has been made (Figure 9-22). First of all, the 2 -way valve has been replaced by a 3-way cam operated valve. This enables the flow controls to be placed so that minimum interactions take place. Also, they are closer to the cylinder, providing better control.

Finally, Figure 9-23 will give the best control of all the pneumatic circuits shown. The flow controls are even more closely coupled, and the flow control interaction is virtually non-existent.

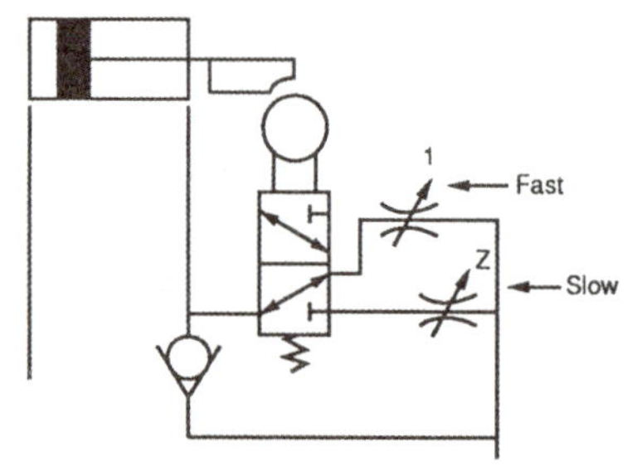

Figure 9-22

SAFETY NOTE: It is extremely difficult to control the speed of a pneumatic cylinder. This is due to the compressibility of air. Therefore each speed control circuit must be examined very closely for all possible failure conditions.

In controlling the cylinder speed there are other problems that must be overcome. One such problem is "jump."

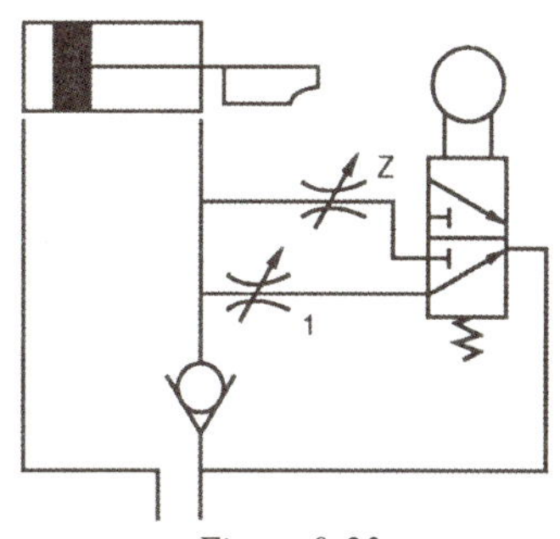

Figure 9-23

Pneumatic Flow Control Problems

In applications, any of the pneumatic controls discussed above may present an objectionable "jump" or rapid partial stroking of the cylinder. This condition may occur when initiating cylinder motion at any position. Supply pressure pushes the piston, which in turn must push out the exhaust air. Until the exhaust air can build up a backpressure the piston will accelerate; once balanced, steady motion should occur. This can occur if the directional control valve is shifted before the cylinder completes its stroke, or if the valve is shifted too soon after the stroke is completed.

Figure 9-24

Quick Exhaust Valves

There are several special-purpose valves that are used in pneumatic circuitry. A quick exhaust valve is an example of one. It is most commonly used in conjunction with a 3- or 4-way directional control valve to increase the exhaust flow (and thus generally increase the rod speed of the cylinder).

The quick exhaust valve typically consists of a resilient disc in a body. The resilient disc "shuttles" side to side, changing the free flow paths through ports in the body.

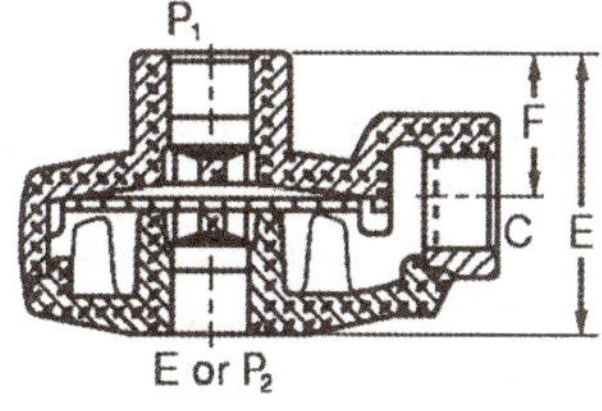

Figure 9-25

How a Quick Exhaust Valve Works

The quick exhaust valve is typically used in conjunction with 3- or 4-way directional control valves. When its inlet port is connected to the supply pressure, it automatically seals its exhaust port and allows pressure to enter the cylinder. If the valve is of the "floppy disc" design, a control valve is shifted so that it can exhaust the supply to the cylinder. The resilient disc moves away from the exhaust port and allows the trapped air in the cylinder to quickly exhaust directly to the atmosphere. This eliminates the need for exhaust air from the cylinder to flow through long, often somewhat restrictive, lines to a main contol valve. Quick exhaust devices should be connected (in juxtaposition) to the cylinder port to give rapid cylinder velocities.

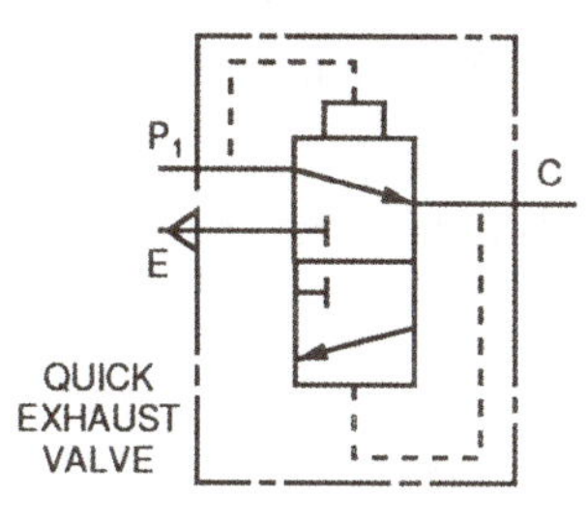

Figure 9-26

A Quick Exhaust Valve in a Circuit

One application for a quick exhaust valve is using single-acting spring return cylinders. With a standard circuit, many times the return speed may be too slow. This may be due to inadequate spring force, high frictional resistance due to the load or the restriction of exhaust flow or a combination of several of these factors. Shown (Figure 9-26) is a single-acting cylinder with a quick exhaust valve closely coupled to the cylinder. This permits greater cylinder speeds than would exist if the air from the cylinder were exhausted through the directional control valve, but without increasing its size.

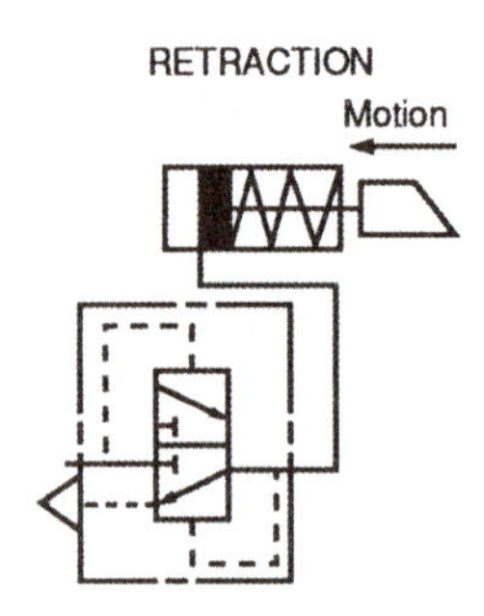

Figure 9-27

During extension of the cylinder, the quick exhaust is part of the supply system that feeds air to the cylinder (Figure 9-27). However, during the return stroke, the quick exhaust shifts as soon as the pressure between it and the directional control valve drops below the pressure at port C. This permits free exhausting of the air from the cylinder directly to the atmosphere, bypassing the directional control valve and all the connecting lines.

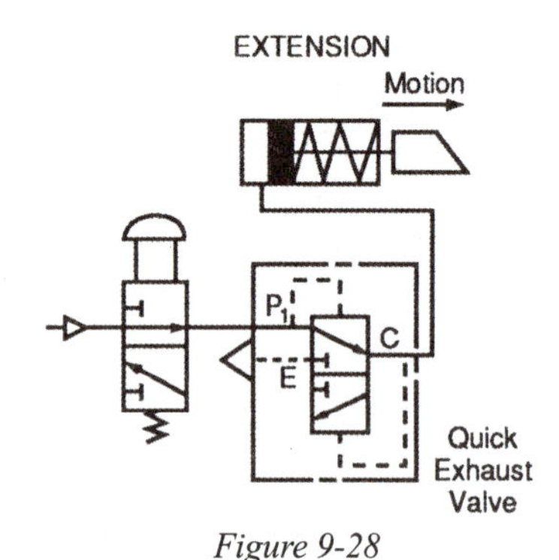

Figure 9-28

Quick exhaust valves may also be added to double-acting cylinder applications. Shown is an application that will deliver an impact blow. In this circuit, a quick exhaust was incorporated at the cap end for high return speed (Figure 9-28). Note that the volume of air under pressure in our example will act as an air spring.

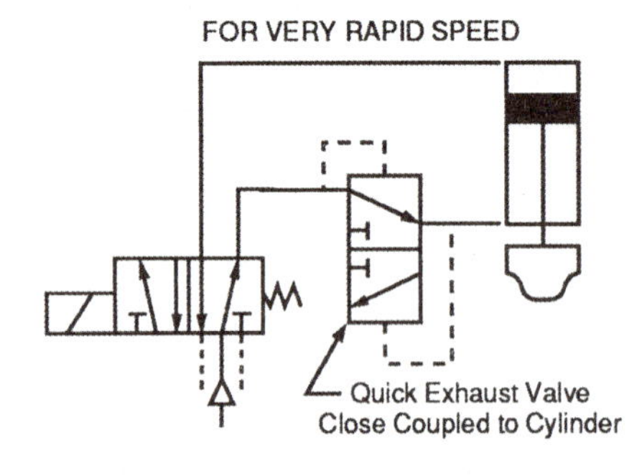

Figure 9-29 NOTE: To insure that the cylinder will deliver an impact blow it should be noted that the cap end of the cylinder must have a pressurized volume. Also, there is no control over the rate of cylinder rod extension.

Quick Exhaust Used as Shuttles

Many quick exhaust valves may also be used as shuttle valves. Shuttle valves are automatic flow path selectors which allow the higher of two pressures to be directed into a flow path. The use of such devices is common in control circuits to "select" the higher sources of pressure within the circuit. However, when exhausting a large volume of air through a large valve, especially through a quick exhaust valve directly to the atmosphere, unwanted air noise may become a problem.

Noise of Expanding Air

Unmuffled exhausting air may generate high intensity sound energy, much of it in the range of normal conversation. Continued exposure to high intensity noises may cause loss of hearing even without noticeable pain or discomfort. Continuous exposure to noise may also cause fatigue, which in turn may reduce operator efficiency, resulting in scrap. Noise may also block out warning signals, thus leading to accidents.

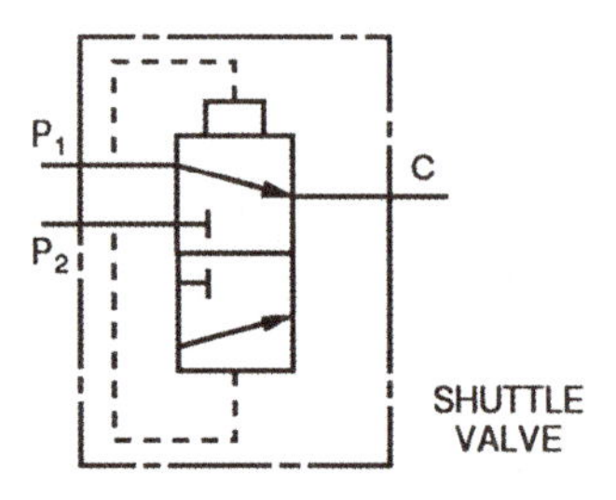

Figure 9-30

Silencers (Mufflers)

Silencer designs range from "tuned" devices such as a volume with a series of baffles, to restricting devices such as porous plastic or bronze.

A good silencer should:

1. have low resistance to flow
2. provide sound attenuation
3. be corrosion resistant

Selecting a Silencer

A silencer should be selected so that the OSHA noise requirements are met. Typically, the max. flow is given a particular sound pressure level. A typical chart is shown in Figure 9-31.

For example, with a 1¼" pipe (32 mm) the maximum flow to atmosphere with 90 dbA is 20 scfm (9.4 dm^3/s). But if the 1¼" pipe size silencer is used, a max. flow of 180 scfm (85 dm^3/s) is obtainable at the same overall sound pressure level. It is also necessary to know the pressure drop through the silencer. This can be estimated from the graph (Figure 9-32). Entering at 180 scfm and moving right, the 1¼" performance line is intersected.

Projecting this point downward, a pressure drop of 3 psi (0.2 bar) is found.

It should be noted that the sound pressure level readings were taken at a distance of one meter (3.28 ft.) from the silencer. Should the distance be different, a correction factor can be applied. This correction factor is shown via a chart (Figure 9-33). The farther away from the source, the more the sound is attenuated or dispersed and thus less intense; moving closer to the silencer, the sound is more concentrated and appears louder or more intense.

Example 9-1

For the silencer selected, what is the db reading at a distance of 3 meters (9.6 ft.) if we consider free field attenuation (i.e. the sound is free to radiate in all directions)?

Solution:
Using the graph and entering at 3 meters and moving upward to the line, a 10 db attenuation is found. This means that the exhausting air will have a sound level reading of 80 db (90-10) at a distance of 3 meters.

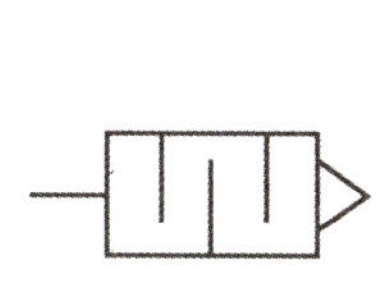

Figure 9-31

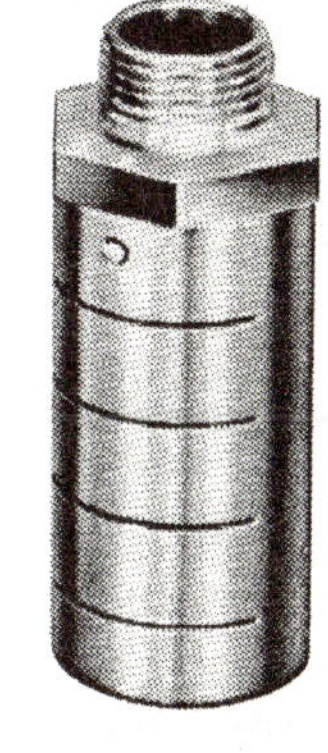

Figure 9-32

Port Size	Max. Flow for 90dB Without "ES" Silencer *	Max. Flow for 90dB With Silencer
¼"	15 scfm	45 scfm
3/8"	5 scfm	70 scfm
½"	5 scfm	110 scfm
¾"	15 scfm	100 scfm
1"	15 scfm	130 scfm
1¼"	20 scfm	180 scfm
1½"	46 scfm	220 scfm

Figure 9-33

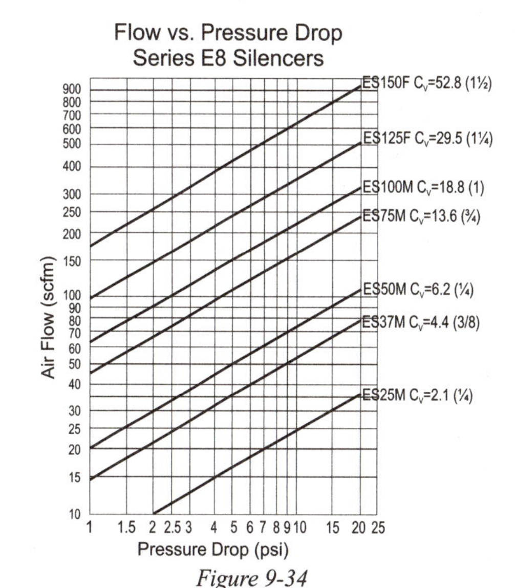

Figure 9-34

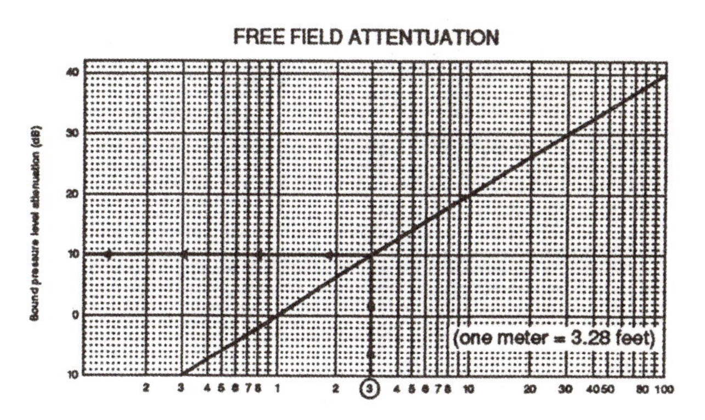

Figure 9-35

Lesson Review

In this lesson dealing with flow controls, quick exhausts and silencers, we have seen that:

- Speed of a pneumatic cylinder is affected by how quickly it can be filled with air at its inlet and exhausted of air at its outlet.
- Flow control valves in pneumatic systems reduce flow rate out an actuator and consequently retard actuator speed by acting as a restriction.
- "Flow control" valves restrict flow by means of an orifice.
- Flow through an orifice is mostly affected by the size of the orifice, and the pressure differential, entrained moisture or oil droplets, contaminants, temperature or the amount of varnish deposited in the restricting area.
- The larger the size of the orifice, the greater the flow rate through the orifice.
- Meter-in flow control circuits control flow into an actuator. These may be used in conjunction with meter-out when jump occurs in some circuit conditions. They will not provide control under all circuit conditions.
- Meter-out flow control circuits control flow discharging from an actuator.
- In restricting air exhausting from an actuator, a flow control valve causes a pressure buildup to occur which adds to the load resistance in order to regulate actuator speed.
- Quick exhaust valves should be used for increasing the speed of a cylinder.
- Quick exhaust valves may sometimes be used as shuttle valves.
- Silencers are devices that are used to combat excessive noise due to exhausting air.

Exercise
Flow Control Valves, Silencers, Quick Exhausts
50 Points

Instructions: Each of the following questions asks the question "What am I?" Identify the components by choosing the most appropriate answer from the selections at the end of the exercise. Each selection is used once.

1. _____ I allow flow to bypass a needle valve.
2. _____ Ball valves, globe valves and needle valves are examples of me.
3. _____ Because of my configuration, flow control valves may be placed in my exhaust port to control forward and/or return speed of a cylinder.
4. _____ In a pneumatic system, a flow control valve may be placed in my exhaust port to control forward and return speed of a cylinder.
5. _____ When exhausting air makes too much noise, I may be used to quiet the problem.
6. _____ In a pneumatic system, I am the best inexpensive configuration which will control the speed of a cylinder.
7. _____ When a high impact speed is necessary, I am the added circuit.
8. _____ When extremely stable velocity control is needed, I will be used.
9. _____ I can allow either of two sources to be directed into a single flow path. I am commonly used in control circuits.
10. _____ On standard pneumatic speed control circuits, I give better control of cylinder speed. What type speed control arrangement am I?

A. 5-ported, 4-way directional control valve

B. Shuttle valve

C. Silencer

D. Adjustable orifice

E. Fixed orifice

F. Quick exhaust valve

G. Check valve

H. 4-ported, 4-way directional control valve

I. Meter-in

J. Meter-out

K. Load resistance

L. Hydraulic dashpot

Chapter 10
Regulators, Excess Flow Valves, Boosters and Sequence Valves

In this lesson pressure control will be discussed, and basic circuits will also be touched upon. The first valve that we will look at is the sequence valve.

Sequence Valve

A normally closed pressure control valve which causes one operation to occur before another is referred to as a sequence valve.

Sequence Valve in a Circuit

In a clamp and form circuit, the clamp cylinder must extend before the form cylinder. The form and clamp cylinders then may retract at the same time. To accomplish this, a sequence valve is positioned in the leg of the circuit just ahead of the form cylinder. The spring in the sequence valve is selected or set so it will not allow the valve to shift to connect primary to secondary passages until pressure is high enough to overcome the set spring force. Then it starts to open and air is allowed to pass from inlet to outlet. In Figure 10-1, the sequence valve setting is 50 psi, which is the minimum pressure needed to perform the job of forming the part.

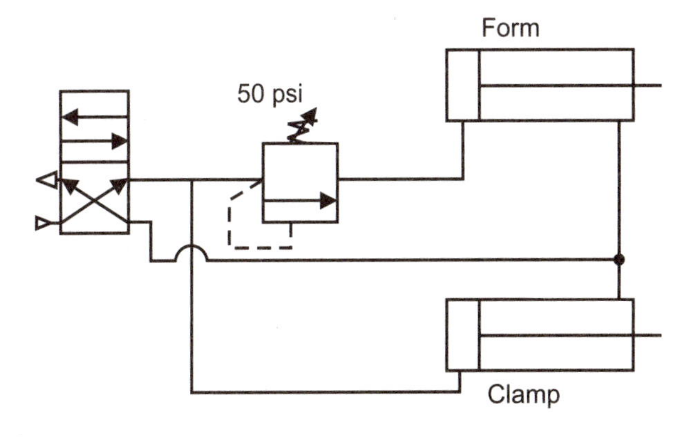

Figure 10-1 ***Safety Note: This circuit has been simplified for instructional purposes. Additional components are required to insure the clamp extending is in place before the form cylinder extends.***

When the directional valve is shifted, flow is directed to the clamp cylinder. Air cannot flow into the form cylinder because the sequence valve is closed. When 50 psi pressure is reached, the sequence valve is actuated, causing primary and secondary passages to be connected. Flow passes to the form cylinder at the pressure needed to perform work, keeping a minimum of 50 psi at the clamp.

If it is important for the form cylinder to retract before the clamp, an additional sequence valve could be added. This is shown in Figure 10-2. Note that in both cases a bypass check has been added around the sequence valve to obtain reverse flow. These sequencing functions could be accomplished with pilot operated valves, either single pilot, spring returned or differential pilot. But getting the air to the point of work is only part of the battle.

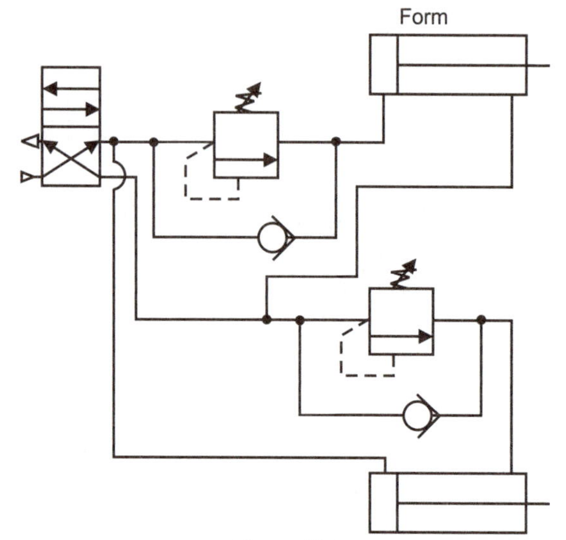

Figure 10-2

Pressure Regulator

The air pressure delivered to the work point is often of a pressure higher than needed. Therefore to control machine forces and to save air, it becomes necessary to reduce the pressure level supplied to the machine and to prevent fluctuation caused by the compressor cycling between its high and low pressure setting.

Depending upon the application, a pressure regulator may have to meet the requirements of:

1. Output (secondary) air at a constant pressure regardless of pressure variation from the upstream pressure and independent flow.
2. Reducing the pressure to a level which meets the work requirement and minimizing the amount of wasted air.

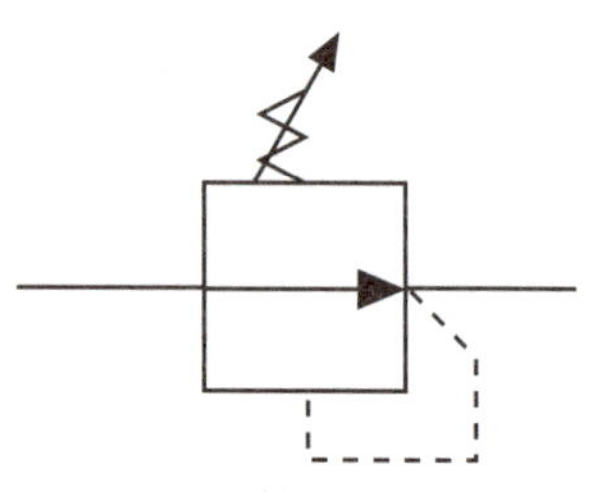

Figure 10-3 Regulator symbol

Acting like a flow control valve, a pressure regulator positioned after an air receiver tank allows compressed air to expand downstream. As pressure downstream rises, it is sensed at the control surface of the regulator through an internal passage at the underside of a piston or diaphragm. This control surface has a large area exposed to downstream pressure and is therefore sensitive to pressure fluctuations. When downstream pressure is high enough to overcome the control force, the control element moves upward, causing the valve to close. The valve, once it seats, shuts off flow and does not allow pressure to continue building downstream. In this way, a relatively constant pressure is made available to an actuator downstream.

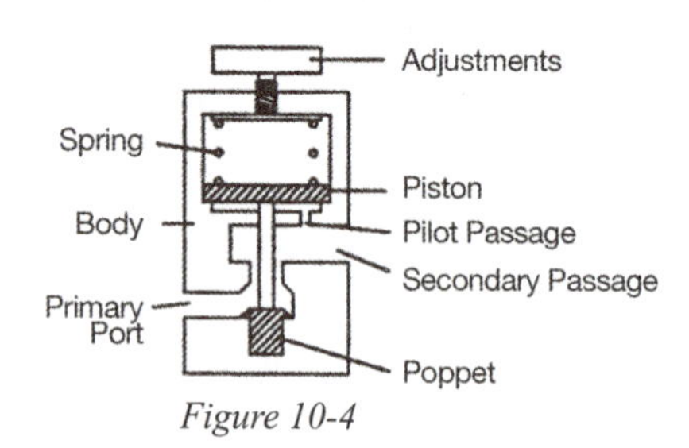

Figure 10-4

Venting Type Regulator

A venting type regulator limits downstream air pressure to a level lower than in the receiver tank, in a manner similar to that described above for a non-venting regulator. It has an additional function in that it acts as a small relief valve for its leg of the circuit in case of any pressure build up in the downstream portion of the circuit.

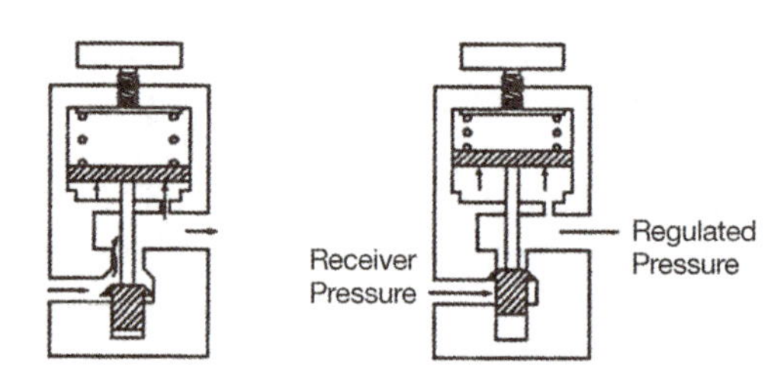

Figure 10-5

Note: The venting or relief flow capability is generally inadequate for critical operations. It is suggested that full flow relief be used whenever large volumes need to be released.

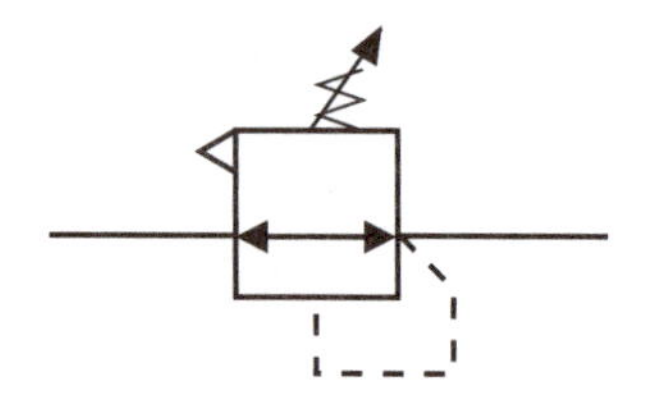

Figure 10-6

What the Venting Type Regulator Consists Of

A typical venting type regulator basically consists of a body with primary and secondary passages, a poppet with a light bias spring, a pressure sensing device with a vent hole, and an adjustable control force, usually a spring. A regulator of this type should have as little restriction to flow as possible in the venting flow direction. This type of flow condition will create a minimum pressure drop across the regulator. A circuit with this type regulator could function without a relief valve. Because of this, savings in labor and materials due to fewer components and connections, would be realized. It should be pointed out that venting, as defined here, is not the same as reverse flow.

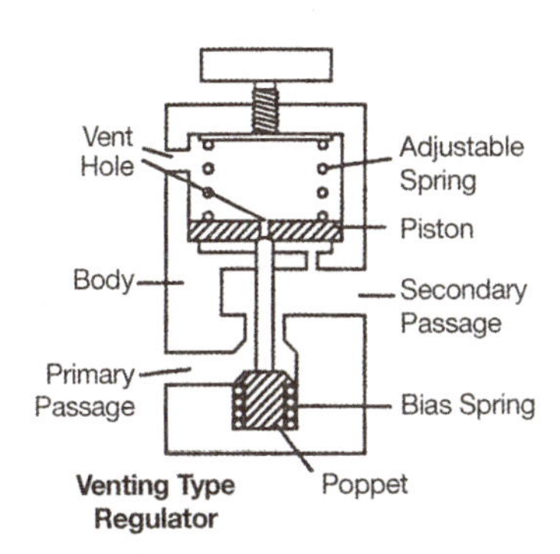

Figure 10-7

Venting type regulators will maintain a set secondary pressure. If for some reason the secondary pressure exceeds a valve setting, a pressure sensing device will open a vent hole reducing the excess secondary pressure to the correct valve.

Reverse flow regulators, unlike many venting regulators, have the capability of passing flow in the opposite direction. This is passing **all** flow, not just enough to maintain a set pressure.

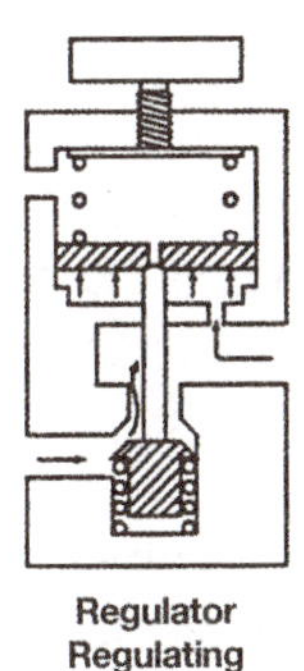

Figure 10-8

How a Venting-Type Regulator Works

A venting-type regulator controls downstream pressure in the same manner as an ordinary regulator while air is being supplied to the system. As pressure rises downstream of the regulator, it is sensed on the underside of the piston or diaphragm. When the force at the piston or diaphragm is large enough to move against the force of the adjustable spring, the piston and poppet move upward, throttling or closing the valve and limiting maximum downstream pressure. The difference in operation between a non-venting and a venting regulator is the venting action taken when a secondary pressure rises above the setting of the regulator. This higher pressure could be the result of a pressure buildup caused by leakage, thermal expansion of air, or a load pushing externally against a cylinder piston. It could also be the result of re-setting the regulator to a lower downstream pressure setting.

When a higher-than-set pressure appears downstream in a venting-type regulator, pressure acts on the piston or diaphragm pushing it up. At the same time, the poppet is pushed up by its light bias spring. With the poppet seated, the piston or diaphragm continues to move until a vent hole is opened. Excess pressure is thus bled off through this hole to the atmosphere.

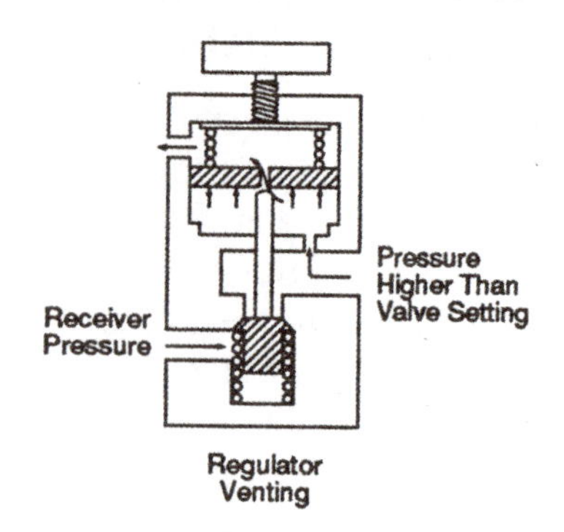

Figure 10-9

When downstream pressure returns to the desired level, the piston or diaphragm reseats on to the poppet stem and pressure regulation resumes. Most pressure regulators found in industrial pneumatic systems are of the venting type.

Diaphragm Regulator

As indicated previously, there is another very common type of pressure regulator. This regulator uses a diaphragm in place of a piston. Operation of a diaphragm regulator is similar to that of a piston type. An example of a venting-type diaphragm regulator is illustrated. Next we should examine a regulator called a pilot controlled regulator.

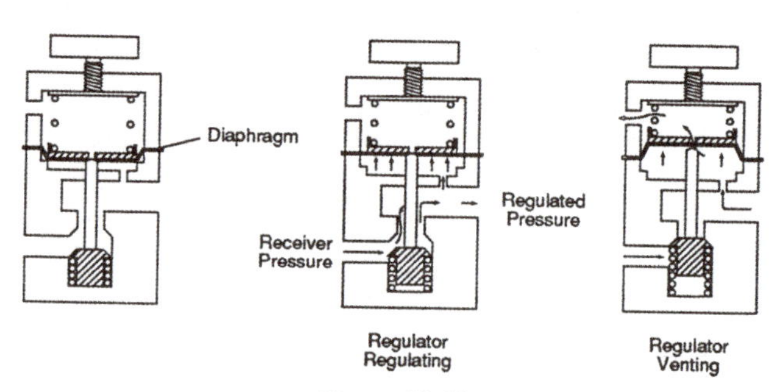

Figure 10-10

Pilot Controlled Regulator

A pilot controlled regulator replaces the adjustable control spring for the principal air flow with air pilot pressure. One advantage of a pilot controlled regulator over a standard regulator is that secondary pressure can be remotely adjusted. Another is that the control force on the control spring side of the piston or diaphragm is almost constant, regardless of the valve position. Because the "spring rate" of the controlling air pressure is constant, a pilot controlled regulator has a distinct control advantage over a spring loaded type which has a "steep" spring rate depending on valve opening, resulting in a range of control pressures dependent on flow rates.

By alternately applying and removing pilot pressure, a pilot controlled regulator can function as a 2-way directional control valve. It can be operated to respond to a series of pre-selected pilot pressures.

What a Pilot Controlled Regulator Consists Of

A pilot controlled regulator basically consists of a valve body with primary and secondary passages, piston or diaphragm, and a poppet biased closed by a light spring.

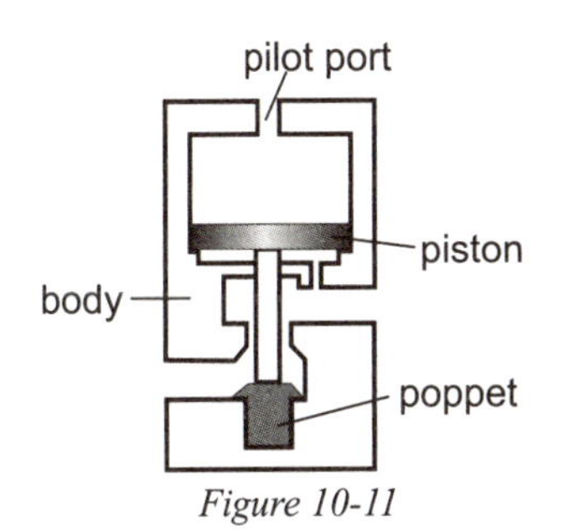

Figure 10-11

How a Pilot Controlled Regulator Works

A pilot controlled regulator operates in the same basic manner as an ordinary regulator. The difference is that pilot air pressure acts against the piston or diaphragm instead of a spring.

As pressure in the downstream circuit of the regulator rises, it is sensed internally to the underside of the piston or diaphragm. When the force generated at the bottom of the piston is greater than the force developed on top of the piston from the pilot pressure, the poppet and piston move toward the poppet seat, regulating flow and thus pressure. Pilot controlled regulators, piston or diaphragm type, can be of the venting or non-venting types and may be designed for reverse flow. However, before we regulate any pressure we must size the regulator.

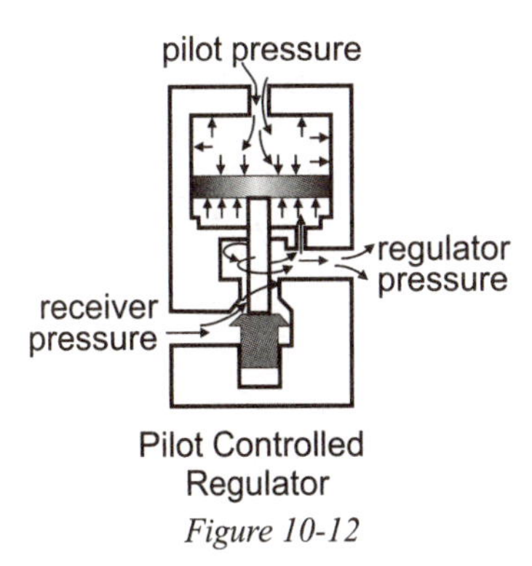

Pilot Controlled Regulator
Figure 10-12

Sizing a Regulator for Flow

A regulator should be sized in such a way that it provides the needed flow rate with a relatively low pressure drop at the desired pressure setting.

The equation for the flow rate for a cylinder application was given in Chapter 6.

For U.S. Units

$$\text{cfm} = \frac{V(\text{in}^3) \times \text{Compression Ratio}}{\text{Time to fill cylinder(s)} \times 28.8}$$

For S.I. Units

$$\text{Flow Rate } (\text{dm}^3/\text{s}) = \frac{V(\text{dm}^3) \times \text{Compression Ratio}}{\text{Time to fill cylinder(s)} \times 28.8}$$

Once flow rates are found, a regulator may be selected. Pressure drop, reverse flow capacity, venting and regulation characteristics may then be checked.

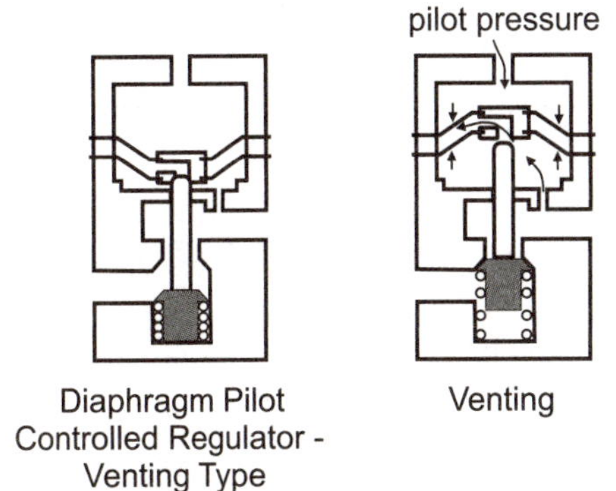

Diaphragm Pilot Controlled Regulator - Venting Type

Venting

Figure 10-13

Example 10-1

A cylinder with a 4" (101 mm) bore, 2" (51 mm) rod and a 20" (508 mm) stroke, extends in 2 seconds and retracts in 1 second. The pressure for extension is 60 psi (4 bar) and 20 psi (1.4 bar) for retraction. What size regulator should be used for this circuit?

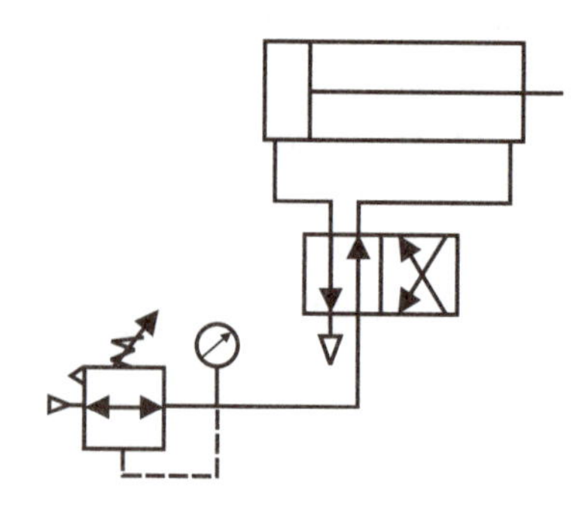

Figure 10-14

Solution:
Since the regulator is controlling the pressure of the air supplied to the directional control valve, both extension and retraction must be considered.

Flow rate for extension

$$\text{cfm} = \frac{\text{V x Compression Ratio}}{\text{time x 28.8}}$$

$$= \frac{(12.56 \times 20) \times \frac{60 + 14.7}{14.7}}{2 \times 28.8}$$

cfm = 22.2 or 10.5 dm³/s)

Flow rate for retraction

$$\text{cfm} = \frac{\frac{\pi (4^2 - 2^2)}{4} \times 20 \times \left[\frac{20 + 14.7}{14.7}\right]}{1 \times 28.8}$$

cfm = 15.4 or 7.3 dm³/s

The regulator will be sized for at least 22.2 cfm. If a basic ¼" ported body with ¼" ports is selected, the following characteristics can be found from catalog data: Figure 10-15 shows the relief and flow characteristics of the regulator. From this graph it can be noted a 5 psi (.34 bar) decrease will occur from the setpoint at a flow of 22 cfm. Also, venting characteristics can be found at the left of the cfm line.

Regulators: Performance Characteristics

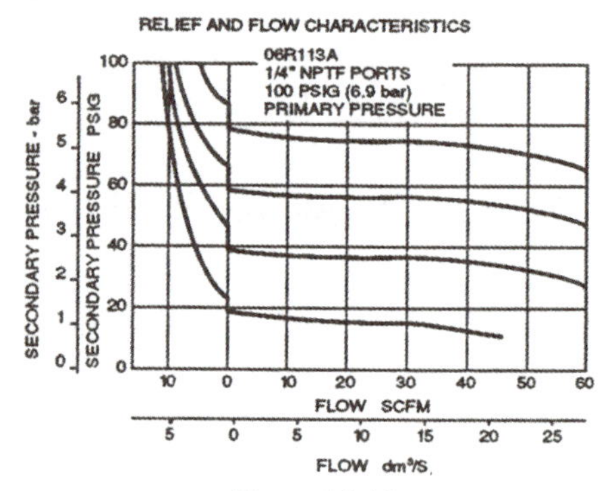

Figure 10-15

Figure 10-16 shows the regulation characteristics of the regulator. It is sometimes referred to as the supply pressure effect. In other words, how much will the secondary (output) pressure vary, with changes in primary pressure? This regulator exhibits excellent characteristics, as noted in the graph.

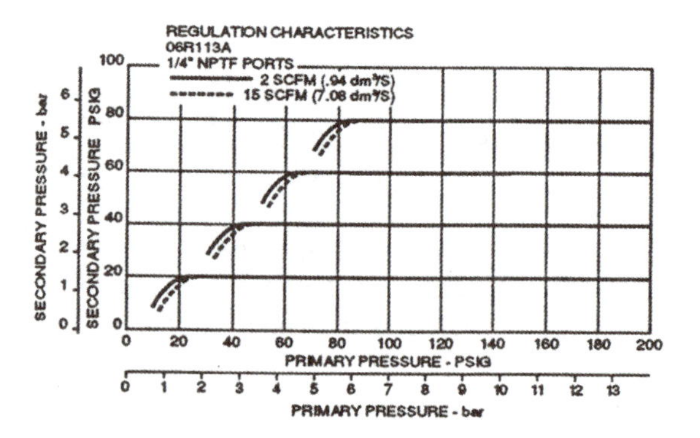

Figure 10-16

A Regulator in a Circuit

As stated before, a regulator has two major functions. One is to pass air at a constant secondary pressure, regardless of the flow variation or upstream pressure and the other is to reduce the pressure to a level that meets with work requirements, minimizing the amount of wasted pressurized air. Let's take a look at the first of these two functions.

The force generated in a pneumatic system is directly proportional to the pressure existing at the actuator. Neglecting frictional forces, the higher the air pressure, the higher the forces. The regulator controls the pressure. In many pneumatic systems a single regulator is used. However, there may exist a lower force requirement for a portion of the cycle. This is illustrated here by a clamp-press application with the desired clamping force being adjustable due to the nature or size of the part being clamped.

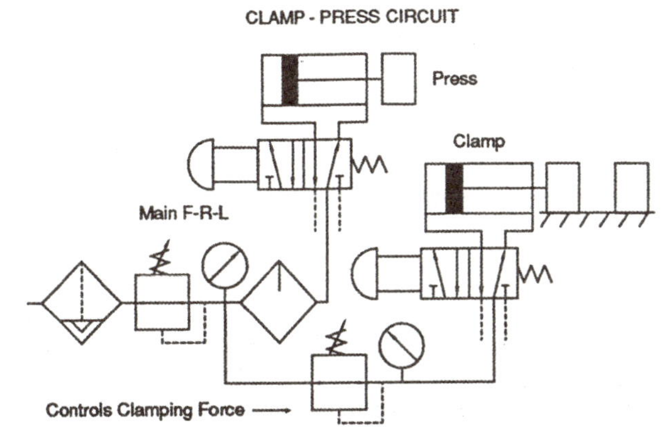

Figure 10-17

By placing an additional regulator downstream of the main regulator, independent control of the clamping force can be accomplished. The main regulator controls the force of the double-acting press cylinder. A procedure such as this could be followed for any multiple force system. Let us now consider function two, energy conservation.

Large amounts of air are wasted when used at pressures higher than are actually required. The higher the pressure of a given volume of air in a cylinder, the greater the consumption of air. In many applications, the pressure used for the "non-work" part of the cycle is the same as the "work" portion. If the "non-work" pressure was less than the required "work" pressure, energy could be saved. There are some methods available to reduce the amount of compressed air used in a circuit, such as differential pressure and dual pressure. First let us look at a differential pressure circuit.

Differential Pressure Circuit

Shown is a differential pressure circuit. The work is performed on the extension stroke while little force, hence a low pressure, is needed to retract the cylinder.

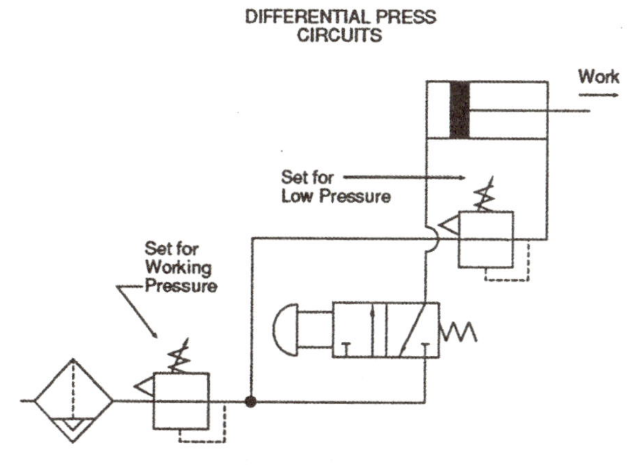

Figure 10-18

A venting type regulator is installed in the head-end cylinder line allowing a considerable savings in air consumption. The cylinder with this arrangement uses reduced pressure acting as an air spring for returning the piston and rod and some external load. Also, notice that a three-way valve may be used since the pressure to the head end of the cylinder chamber remains constant (Figure 10-18).

Dual Pressure Circuit

Another configuration for obtaining dual pressure is shown in figure 10-19. It utilizes the principle that many types of 5-ported 4-way valves may be used with either dual pressure or dual exhaust. Actuation of one solenoid allows one pressure to enter one end of the cylinder.

Actuation of the other solenoid reverses the action as it allows a second pressure to enter the other end of the cylinder (Figure 10-19).

The following example will show the potential savings.

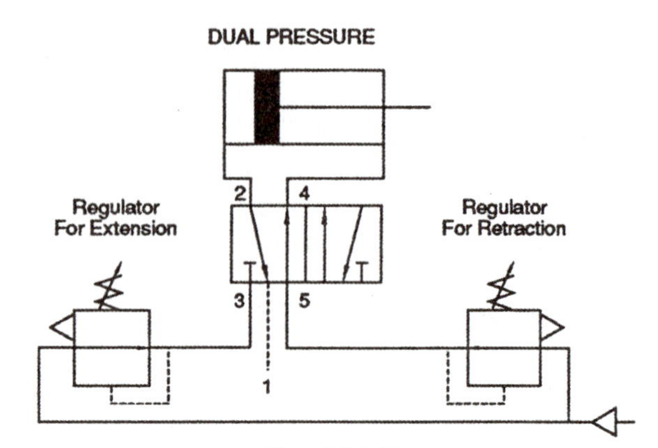

Figure 10-19

Example 10-2

A cylinder needs a pressure of 65 psi (5 bar) to do work during extension. The retraction stroke needs only 25 psi (2 bar). What is the difference in the cost per year between running a single regulator circuit and (10-19). Specifications - 5" (127mm) bore; 2" (51 mm) rod; 18" (457 mm) stroke; extended time - 2 sec; retract time - 1 sec; circuit runs 7 hours a day, 6 days a week, 50 weeks a year.

Solution:
First we need to calculate the flow rate in to the cylinder from Chapter 7.

$$\text{cfm} = \frac{V(\text{in}^3) \times \text{Compression Ratio}}{\text{Time to fill cylinder(s)} \times 28.8}$$

The area of the piston is:

$$Ap = \pi/4\,(5)^2 = 19.63 \text{ in}^2 \ (126 \text{ cm}^2)$$

$$Ae = \pi/4\,(5^2 - 2^2) = 16.5 \text{ in}^2 \ (106 \text{ cm}^2)$$

The volumes are:

$$V_{(extension)} = 19.63 \times 18 = 353 \text{ in}^3 \ (5.76 \text{ dm}^3)$$

$$V_{(retraction)} = 16.5 \times 18 = 279 \text{ in}^3 \ (4.85 \text{ dm}^3)$$

The compression ratios are:

$$\text{Compression Ratio}_{(65)} = \frac{65 + 14.7}{14.7} = 5.42$$

$$\text{Compression Ratio}_{(25)} = \frac{25 + 14.7}{14.7} = 2.7$$

The flow rates for Figure 10-18 are:

Flow rate extension =

$$\frac{353 \times 5.42}{2 \times 28.8} = 33.2 \text{ cfm } (15.7 \text{ dm}^3/\text{s})$$

Flow retraction =

$$\frac{297 \times 5.42}{1 \times 28.8} = 55.9 \text{ cfm } (26.4 \text{ dm}^3/\text{s})$$

$$\text{Average flow rate} = \frac{(33.2 \text{ cfm} \times 2) + (55.9 \text{ cfm} \times 1)}{(2 + 1)}$$

$$= 40.8\ \text{cfm}_{average}\ (19.3 \text{ dm}^3/\text{s})$$

The flow rates for Figure 10-19 are:

Flow rate extension =

$$\frac{353 \text{ x } 5.42}{2 \text{ x } 28.8} = 33.2 \text{ cfm } (15.7 \text{ dm}^3/\text{s})$$

Flow rate retraction =

$$\frac{297 \text{ x } 2.7}{1 \text{ x } 28.8} = 27.8 \text{ cfm } (13.1 \text{ dm}^3/\text{s})$$

Average flow rate =

$$\frac{(33.2 \text{ x } 2) + (27.8 \text{ x } 1)}{(2 + 1)} = 31.4 \text{ cfm } (14.8 \text{ dm}^3/\text{s})$$

Flow rate per year for Figure 10-18:

= 40.8 x 60 x 7 x 6 x 50

= 5,140,800 cfm/year (2426 kdm³/yr)

Cost of air is approximately $0.15 per 1,000 ft³.

Cost of air (10-18)

= 5,140,800 x 0.15/1000

= $771

Flow rate per year for 10-19

= 31.4 x 60 x 7 x 6 x 50

= 3,956,400 cfm/yr (1867 kdm³/yr)

Cost of 10-19

= 3,956,400 x 0.15/1000

= $593

Savings

= 771 - 593

= $178/year (per year/per cylinder)

Boosters

Sometimes the pressure needed for a particular part of a cycle is in excess of the plant's air supply pressure. Examples of such conditions would be die cushion cylinders, test stands and the like. If the volume of air needed is used in pulses and the needed volume is not large, a reciprocating booster may be the economical answer. Boosters may be of the air-to-air or air-to-oil types. They also may be single or dual pressure. Single pressure will deliver compressed fluid from the intensifier. Dual pressure first delivers pressure from the main system to pressurize high pressure fluid from the intensifier.

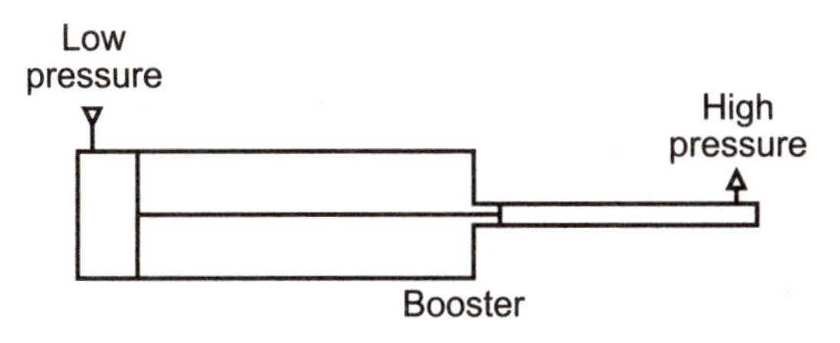

Figure 10-20

No matter what type of intensifier is selected, certain guidelines should be followed:

1. Intensifiers are generally faster operating when:
   - There is adequate input pressure.
   - The ports and piping are not oversized. Consider the use of ports and connecting lines that reduce the volume necessary to compress.
   - The intensifier draining cylinder is pre-exhausted and the high pressure (driven) cylinder is at line pressure before the power stroke occurs.
2. Bypass the intensifier with a pre-fill low pressure line by direct connection through a check valve to the pressure vessel and similarly to a dual pressure intensifier.
3. Regulate the driving pressure to the intensifier to achieve the required high pressure output.
4. Keep all piping lengths beyond the control valve to a minimum by having the banks, intensifier and pressure vessel as close together as possible.
5. A single pressure intensifier usually provides faster cylinder action because it does not need to change from low to high pressure, but instead immediately supplies the high pressure.
6. Intensifiers are generally used in circuits where limited quantities of high pressure fluids are required.

Air-Oil Tanks

One type of intensifier is the air-oil tank. Air-oil tanks accept compressed gas and transfer the pressure directly to a liquid for use as the control medium. Since hydraulic fluid is more easily controlled (flow), a smooth and accurately controlled flow rate can be obtained. Typically air-oil tanks have a sight level gage to show the oil level. They also contain a series of baffles. The upper disperses the incoming air over the surface of the oil in such a way to minimize agitation and aeration.

The lower baffle insures a smooth flow pattern of the liquid that minimizes oil turbulence and eliminates swirling, funneling or splashing which in turn could cause oil aeration or the oil to be blown from the tank into the exhaust air.

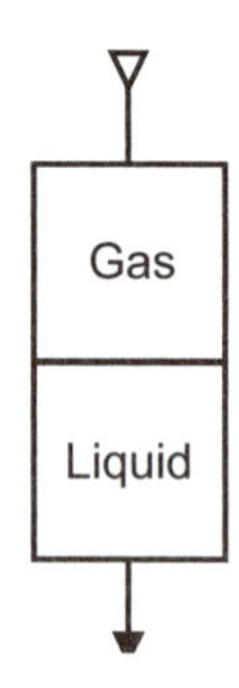

Figure 10-21

Air-to-Air Booster Circuits

This is an air-to-air circuit of the single pressure type. It consists of a double-acting air-to-air booster (1), a high pressure receiver (2) and a cycling circuit. This type of system is a demand type, functioning only to bring the receiver pressure up to the desired pressure, which is controlled by regulator (3). Control valve (4) is the device that cycles the booster, controlled by the limit valves (5) and (6). Thus booster motion continues until a force balance exists. After this happens, the rate of reciprocation is determined by system leakage (Figure 10-22).

Note: To help insure the correct circuit operation, the cam operated valves should be snap acting or there should be rapid traverse of the intensifier (1).

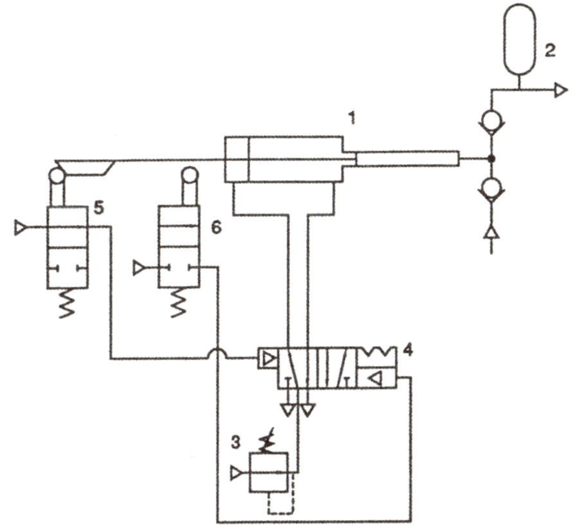

Figure 10-22

Air-to-Oil Booster Circuits

Now if a relatively low volume of high pressure fluid is needed in an air system, an air-to-oil booster can be used. Typical examples may be clamping, pressing, stamping or holding molds with high forces for long periods of time. These may also be single to dual pressure types.

A single pressure type is shown in Figure 10-23. It consists of an intensifier, air-oil tank, flow control valve and single-acting cylinder. This configuration is commonly called a closed system because the fluid necessary for stroking the cylinder is supplied only from the intensifier. The flow control determines the rate of unclamping. The clamping force is controlled by the pressure regulator.

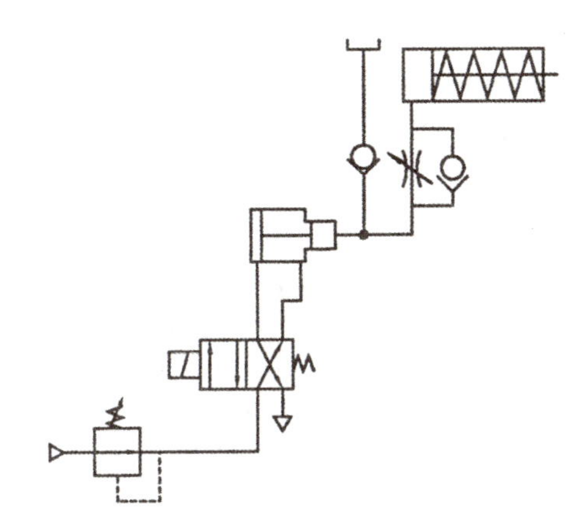

Figure 10-23

Figure 10-24 is the basic circuit for a dual pressure system supplying pressure to a double-acting work system. The input pressure is introduced to the system through shop air lines to the 4-way directional control valve "A". When valve "A" is shifting to position as shown, air is directed into air-oil tank "B" and to valve "C". Oil acted upon by air pressure is forced from the air-oil tank through the pressure chamber of the retracted intensifier and into the work cylinder. The cylinder advances in stroke, being driven by this incoming oil. At a predetermined point in the stroke length of the work cylinder, valve "C" is synchronized to shift and direct air pressure to the intensifier to drive it in its power stroke, isolating tank "B" and supplying high pressure to the work cylinder for its high thrust stroke. The work cylinder and intensifier are retracted by shifting valves "A" and "C" to exhaust the intensifier and tank "B". At the same time, air pressure is directed to tank "D" and

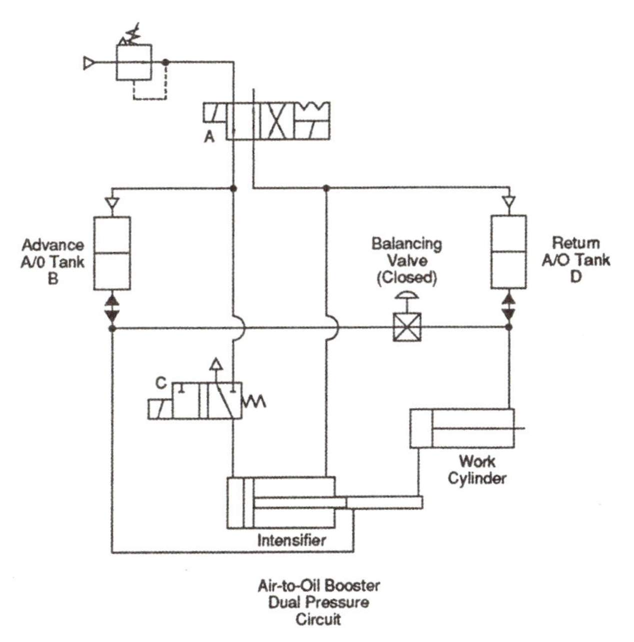

Air-to-Oil Booster
Dual Pressure
Circuit

Figure 10-24

to the rod end side of the intensifier piston. Oil from tank "D" retracts the cylinder at pressure. This is an over simplified circuit. Let us now look at a device that is required should one of our air lines break. That device is the excess flow valve and air fuse.

Excess Flow Valves

The Occupational Safety and Health regulation for Construction Par. 1926.302, dated April 6, 1979, requires all air hose systems larger than ½" be equipped with a safety device to reduce pressure in the event of hose or fitting failure. This is intended to minimize the probability of hose whipping if the hose breaks. The valves used in these applications are called air fuses or excess flow valves. The valve is normally open type that automatically closes when the flow rate suddenly increases past a predetermined rate, such as occurs when a line breaks.

One design is in Figure 10-25. A poppet is held in the open position by a spring. When a line breaks causing air flow to suddenly increase beyond a certain preset limit, the surge of air drives the poppet to its closed position. When the break is repaired and pressure equalized, the spring returns the poppet to its normal position. This valve is required off the air tank and always one per hose line of 100 feet (31 m) or less. One additional valve is required for every 100 feet (31m) after the initial 100 foot run. Also, one safety valve is required per outlet on a manifold and on each branch of a "Y' connection of a hose.

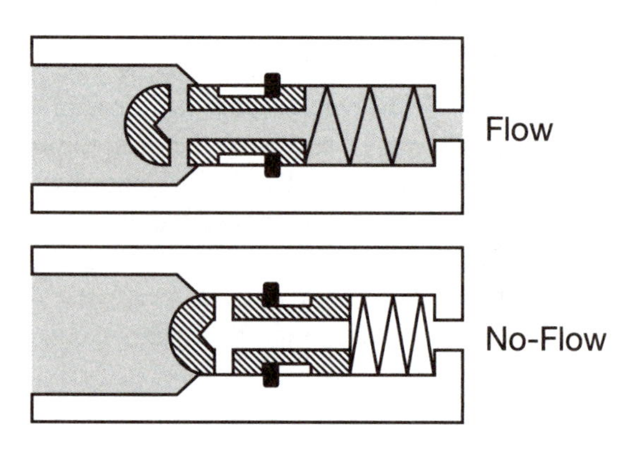

Figure 10-25

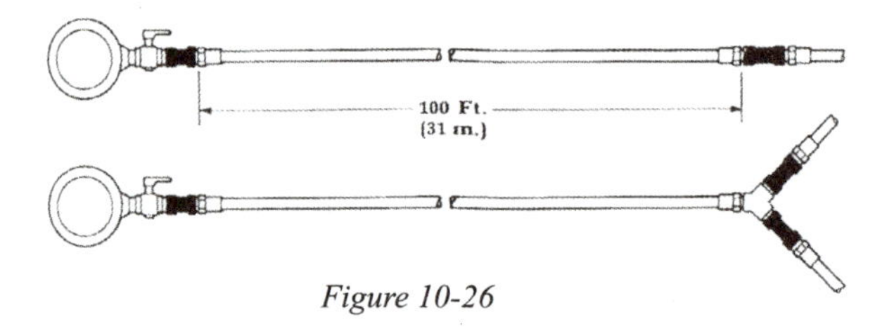

Figure 10-26

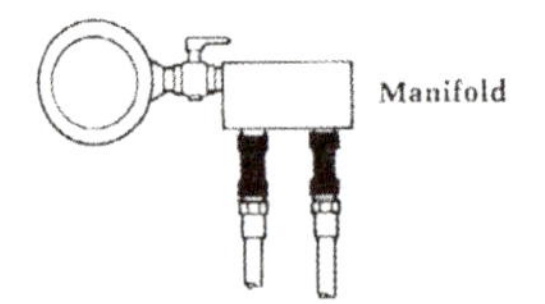

Figure 10-27

Lesson Review

In this lesson dealing with pressure controls we have seen that:

- Simple pressure controls usually (typically, generally, etc.) have their moveable parts biased by spring pressure only.
- A sequence valve is a normally closed pressure valve which may be used for control of one operation to occur before another by blocking air passages until a set pressure is reached.
- Pressure regulators, in an industrial pneumatic system, are most often venting type rather than non-venting.
- A piston type regulator is a common type of pressure regulator found in an industrial pneumatic system.
- One advantage of a pilot controlled regulator over a simple regulator is that regulation pressure can readily be remotely adjusted and programmed.
- A pilot controlled regulator replaces the adjustable control spring in the power stage with air pilot pressure.
- Regulators should be sized for the maximum flow rate in the system.
- Use dual regulation for energy savings.
- Boosters are used to obtain a high pressure in a circuit where the volume needed is not large.
- Boosters may be air-to-air or air-to-oil types.
- Air-oil tanks are provided to accept compressed air and transfer the pressure to a liquid for use as the control medium.
- Excess flow valves are used to reduce pressure in the event of hose or fitting failure.

Exercise
Regulators, Excess Flow Valves, Boosters and Sequence Valves
60 points

Complete the following problems :

1. A single-acting cylinder extends in 0.5 seconds. Size the regulator (cfm) for the circuit.

   Design Criteria: 8" (203 mm) bore
   4" (102 mm) rod
   14" (356 mm) stroke
   pressure to move load, etc. 55 psi (4 bar).

2. What are the yearly cost saving comparing the following two circuits?

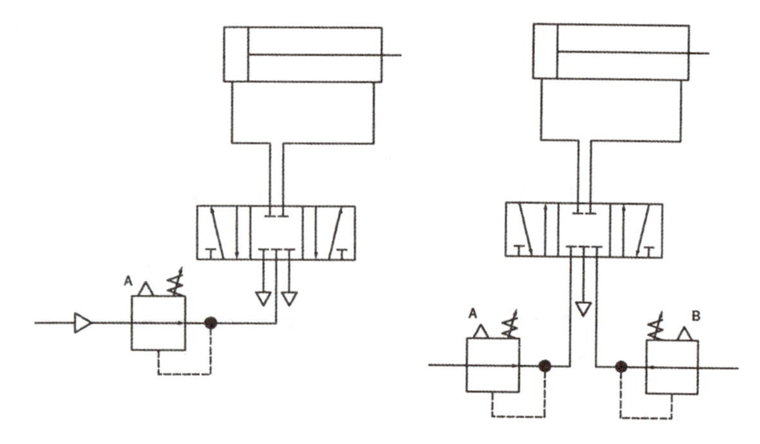

   Design Specifications: 6" (152 mm) bore
   4" (101 mm) rod
   12" (305 mm) stroke
   Extension time (0.5s)
   Retraction time (0.25s)
   Cost of air is \$0.15/1000 ft³
   Machine runs 16 hours a day, 6 days a week, 50 weeks a year
   Regulator A is 70 psi (5 bar), Regulator B is 20 psi (1.4 bar).

3. What are the yearly cost savings comparing the two circuits?
   (Use figures in Question 2 with the following design specifications.)

   Design specifications: 14" (355 mm) bore
   5" (127 mm) rod
   30" (762 mm) stroke
   Extension time 2s
   Retraction time 2s
   Cost of Air is \$ 0.15/1000 ft³.
   Regulator A is 80 psi (6 bar), Regulator B is 20 (1.4 bar)
   Machine runs 16 hours a day, 6 days a week, 50 weeks a year

Chapter 11
Air Preparation

Pneumatic systems and components discussed up to this point require compressed air free of contamination. No matter how well a system is designed or how expansive or sophisticated a particular component may be, contaminated air will interfere with components and system operation. In general, the more contaminated the air, the less dependable a pneumatic system will be.

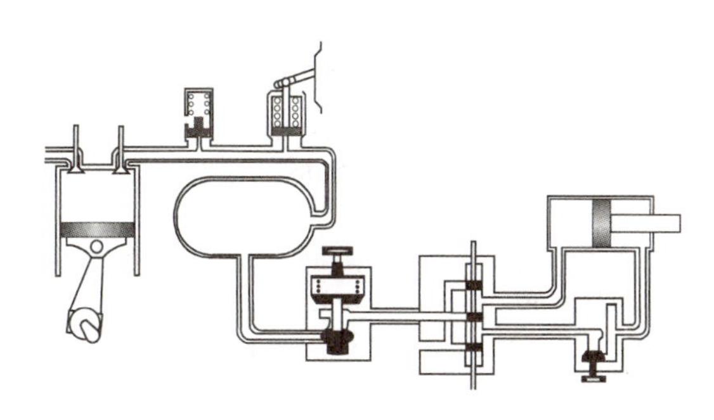

Figure 11-1 A simple pneumatic system

Even the air we breathe may not be appropriate for a specific pneumatic system. Air must be conditioned; it must be decontaminated before it is used in the pneumatic system.

Contaminants in a pneumatic system come from three basic sources: built-in, generated and ingested.

"Built-in" dirt occurs in newly fabricated systems where components or piping are dirty or where installation practices are below standard.

As a system is assembled, pipes, valves and storage tanks become a collector for rust, paint chips, dust, sealant tape, cigarette butts and grit. Many harmful dirt particles are invisible to the unaided eye and cannot be removed by wiping with a rag or blowing off with an air hose.

The second source of dirt is that generated within the working system itself. As a system operates, moving parts in contact with other surfaces naturally begin to wear, generating wear particles. The use of incorrect types of fluid conductors may cause rust flakes or particles to form. These are eventually carried down through lines to tool stations.

The third source of contaminants are those added to the system. If a valve breaks down, the maintenance man may replace the component or repair it on the spot. In either case, he will more than likely be working in a dirty environment, which may allow contamination of the system as soon as a line is disassembled.

Dirt can also be added to a system by means of a cylinder. After a time, a cylinder rod wiper wears its outer sealing edge so it can no longer wipe off fine particles of dirt. In this condition, dirt may be drawn into the system each time the cylinder is stroked. Quick disconnects are another potential source for externally generated contaminants.

Contaminant Sources:
Built-in
Generated
Ingested

Contaminant Type

The contaminants found in a system may be divided into three groups: abrasive dirt, soft dirt and entrained liquids.

Hard dirt may come from inside or outside the plant. Hard dust, grinding compounds, and foundry sand are just a few examples. This dirt is abrasive, affecting the proper operation of components. With this type of component, dirt can wedge into clearances between moving parts causing faulty operation of components.

Figure 11-2 Types of contamination

Lastly, entrained liquids usually enter the pneumatic system through entrainment in the air. In large quantities, moisture can wash away lubrication and, in any quantity, will cause rust to form. Also, oil carried over from top end compressor lubrication can cause resilient seals to deteriorate. This is especially true if synthetic lubricants are used with standard seals. Without effective seals, a pneumatic system wastes its stored energy. Also, operational problems may occur because seals become swollen, making valve shifting erratic.

Now that we know what type of contamination we may be dealing with, what about size?

To measure contaminants we use the micrometer scale.

The Micrometer Scale

One micrometer is equal to one millionth of a meter or thirty-nine millionths of an inch. A single micrometer is invisible to the naked eye and is so small that it is difficult to imagine. To bring the size more down to earth, some everyday objects will be measured using the micrometer scale. For example, an ordinary grain of table salt measures 100 micrometers (µm), and the average diameter of human hair measures 70 micrometers (µm). for comparison, twenty-five micrometers is approximately one thousandth of an inch.

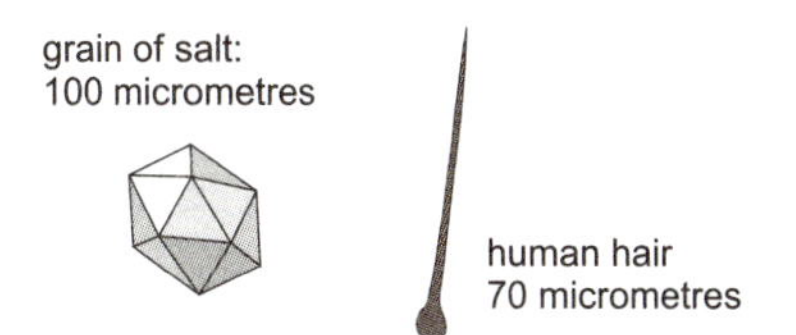

Figure 11-3

Relative Sizes of Particles

Substance	Microns	Inches
Grain of table salt	100	.0039
Human hair	70	.0027
Lower limit of visibility	40	.0016
Milled flour	25	.0010
Red blood cells	8	.0003
Bacteria	2	.0001

Figure 11-4

Limit of Visibility

The lower limit of visibility for an unaided human eye is 40 micrometers. In other words, the average person can see individual particles measuring 40 micrometers and larger.

However many of the harmful dirt particles in a pneumatic system are below 40 micrometers.

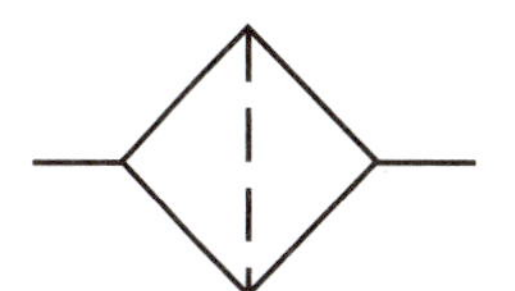

Figure 11-5

Air Line Filter

As previously discussed in the chapter on compressors, the first line of defense for industrial compressed air is the compressor intake filter and the aftercooler.

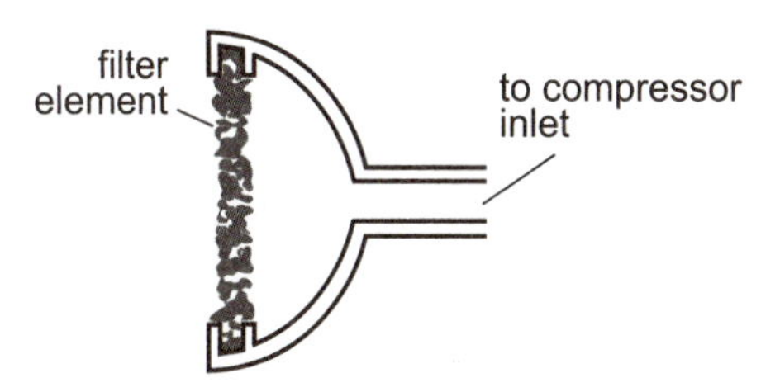

Figure 11-6

Proceeding from the aftercooler the air enters a receiver tank where further cooling (therefore a reduction in moisture content) takes place. Additional moisture may be removed as the air preparation takes place at the individual pieces of equipment that are served by the plant's compressed air system. An air line filter is a device that is placed in the air line at the work station to be protected. By doing this, it removes particulate matter from all of the air to pass through the filter element.

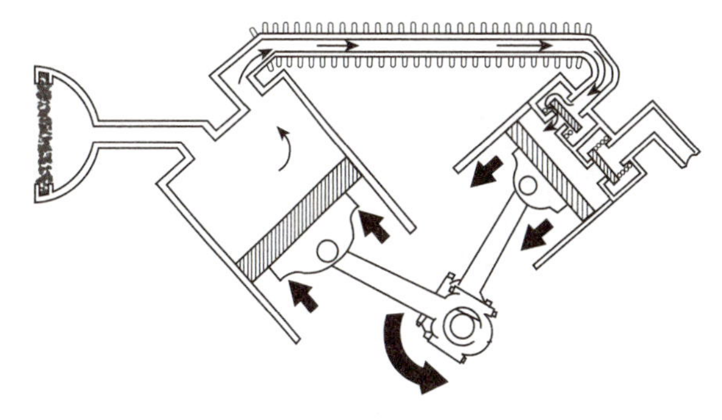

Figure 11-7

What an Air Line Filter Consists Of

An air line filter basically consists of a housing with inlet and outlet ports, deflector, shroud, filter element, baffle plate, filter bowl and drain.

How an Air Line Filter Works

Filtration through an air line filter takes place in two stages. In the first stage, air enters the inlet port and flows through the openings in the deflector plate (A) which causes a swirling action. Entrained liquids and large (heavy) particles are forced out to the bowl's interior wall (B) by the centrifugal action of the swirling air.

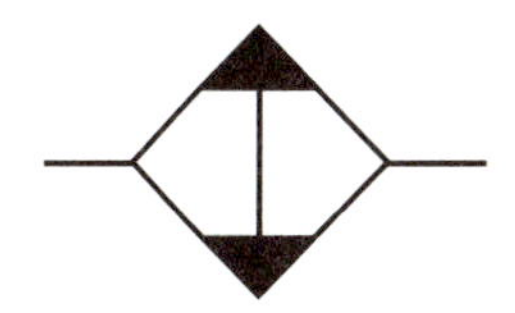

Figure 11-8

They then run down the sides of the bowl. Shroud (C) assures that swirling action occurs at low flow rates and uniformly distributes the air flow across the entire length of the elements (D). It also prevents debris from concentrating at any one point on the element.

The baffle (E) separates the lower portion of the bowl into a "quiet zone" where the removed liquid and particles collect unaffected by the swirling air and are not re-entrained into the flowing air.

In the second stage of filtration, air flows through the element (D) where smaller particles which are still airborne are filtered out and retained. The filtered air then passes downstream.

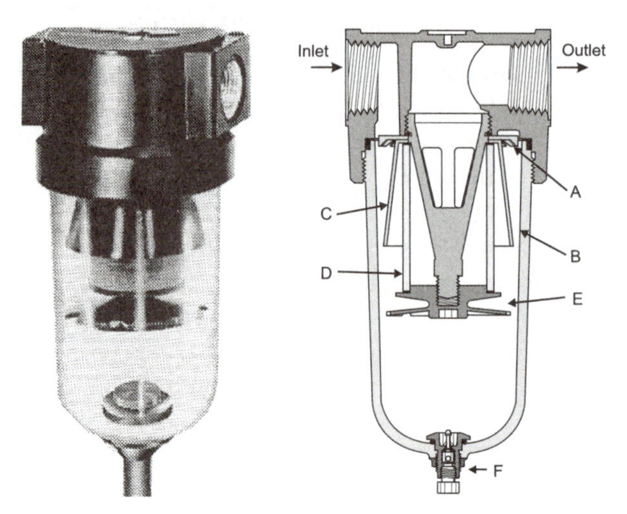

Figure 11-9

Filter Elements

The function of the mechanical filter element in a pneumatic system is to remove particles of dirt from the air system. This is done by forcing air flow to pass through porous materials. Typical filter materials include cloth, felt, porous sintered metal, wire mesh, ceramics, polyethylene, plastics and resin-impregnated paper. Some are throw away types while others may be cleaned and reused. Filter elements are generally divided into two types, depth and edge.

Figure 11-10 Depth type element

Depth Type Elements

With a depth type element air is forced to pass through an appreciable material thickness. Dirt is trapped because of the tortuous path the air must take. Depth elements in pneumatic systems are frequently made of porous bronze or plastic.

Edge Type Elements

Conversely, edge type elements offer an air stream, a relatively straight flow path. Dirt is caught on the surface or edge of the element which faces the air flow. Edge type elements in pneumatic systems are often made of resin-impregnated paper ribbon.

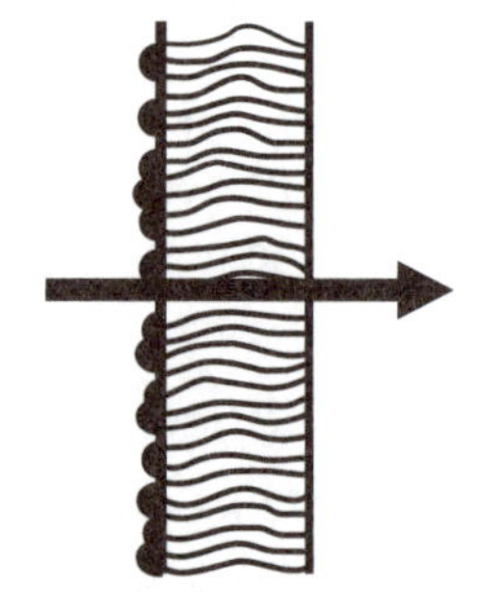

Figure 11-11 Edge type element

Pore Size Air Filter Elements

Because of its construction, an air line filter element may have many pores of various sizes. Many of the pores are small. A few pores are relatively large. If it has no consistent hole or pore size, an air line filter element is given a nominal rating.

Nominal Rating

A nominal rating is an element rating given by the filter manufacturer, indicating the expected average hole size in the element. For example, a depth element with a nominal rating of 40 micrometers indicates that the majority of the pores in the element are 40 micrometers in size and that a large percentage of the 40 micrometer particles will be trapped.

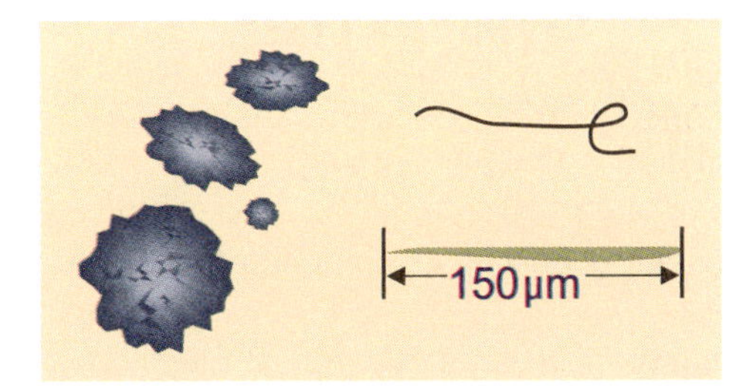

Figure 11-12

Filter Ratings in Practice

Air line filter elements found in industrial pneumatic systems generally have nominal ratings ranging from 50 to 5 micrometers. Since dirt in a pneumatic system comes in all sizes, shapes and materials, no guarantee is made as to what size particles will be removed from a compressed air stream. Now that we have removed the contaminants we must remove the liquids. This can be done with an oil removal filter.

Oil Removal Filter

Oil removal filters are a relatively recent development and are designed to meet the need for delivering oil-free air for a number of industries. A partial list of the industries requiring this type of filtration would include:

1. Dental and medical
2. Chemical
3. Pharmaceutical
4. Bottling and packing
5. Food processing
6. Many types of industrial paint spraying

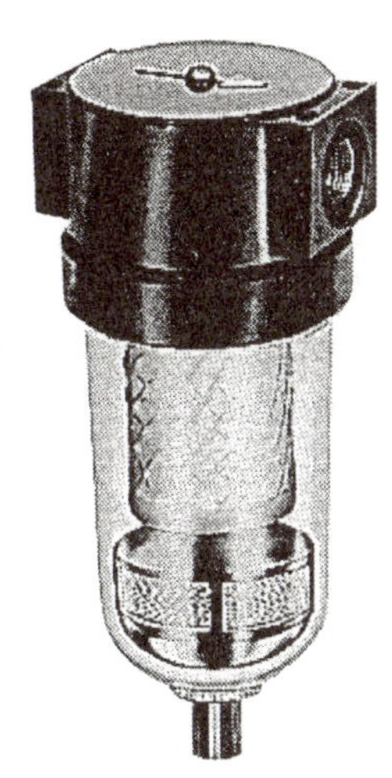

Figure 11-13

Specific applications include pneumatic control instrumentation, pneumatic measuring and gaging, protection of air bearings, paint spraying, blow molding machinery, moving part logic devices, printing processes, paper separation, production and packaging of fine chemicals, photographic process, absorption air dryers or any place where oil type compressors are used but oil free air is required. The features of such a device are that it removes liquid aerosols and submicronic particles down to approximately 0.3 micrometers. The efficiency of removing such liquid aerosols is as high as 99.9%. Usually a prefilter is added to increase life and efficiency.

How an Oil Removal Filter Works

The contaminated air enters the element interior and is forced through a membrane thickness of borosilicate glass fibers, sometimes coated with epoxy. The flow then passes through an outer structural support, and at this state most of the sub-micrometer oil particles carried by the air have been removed. The droplets coalesce and are blotted from the filter surface by layers of non-woven glass felt and rayon cloth. The drops can gravitate to the filter sump where they may be drained periodically.

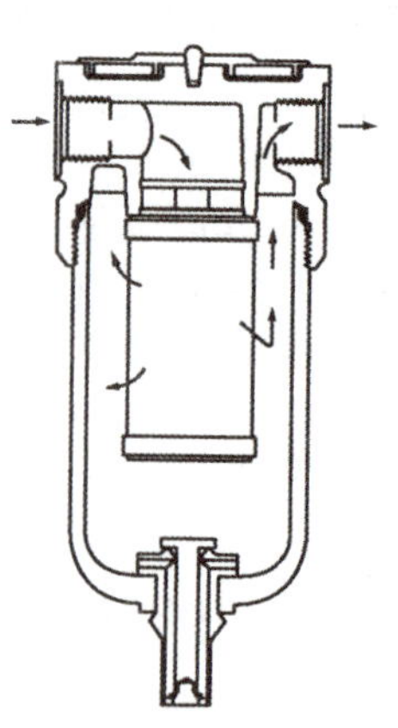

Figure 11-14

Draining a Filter

The liquids and particles collected in the quiet zone must be drained before their level reaches a height where they would be re-entrained into the air. This can be accomplished manually or with an automatic drain.

The manual drain is standard on most filters. This type requires maintenance personnel to periodically check each unit and open the drain if contaminant level is high. These drains may consist of a petcock type, similar to a car radiator drain cock, or other manually operated types which are of a spring loaded design which must be pressed to open.

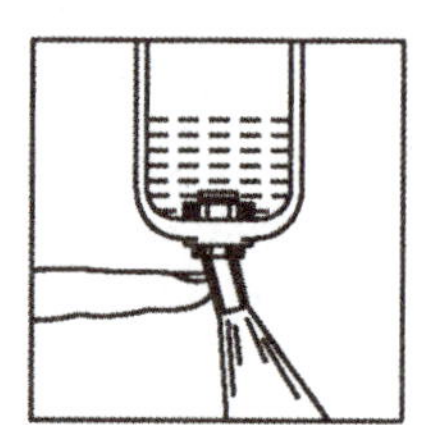

Figure 11-15

Automatic Drain

Another type of drain, the automatic drain, is a device placed in the quiet zone of the filter's bowl. It typically has a float that when raised, directly or through a pilot action, causes a drain port to open, expelling liquid and entrained contaminants.

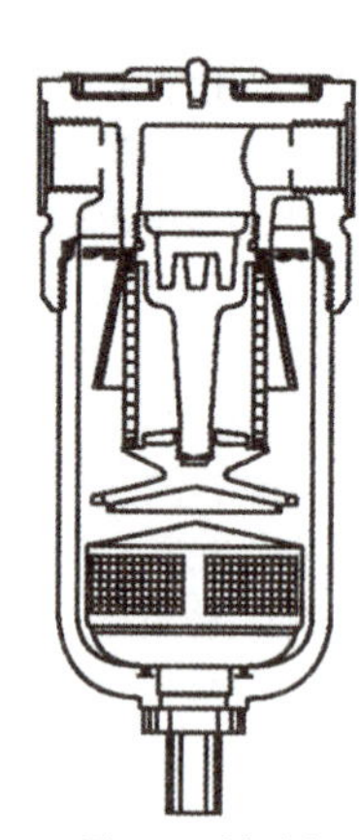

Figure 11-16

How An Automatic Drain Works

When system pressure is applied, it enters under the diaphragm and lifts it and the spool upward, sealing the spool nose to the bleed seal. The bleed seal and pin continue to rise, lifting the float and float gasket from the top body orifice. Pressure begins to enter above the diaphragm causing it to reverse and move down. At some point in the downward motion, the bleed seal and nose separate after the float/float gasket have sealed the top body orifice. Thus, air trapped in the chamber above the diaphragm can vent through the spool and become slightly decreased in pressure.

When the forces above and below the diaphragm come into balance, the diaphragm, spool and pin have moved to a position allowing the float gasket to form a seal at the top body orifice and the spool nose has closed the bleed seal. Pressure above the diaphragm remains slightly less than below.

Automatic Filter Drain (Closed Position)

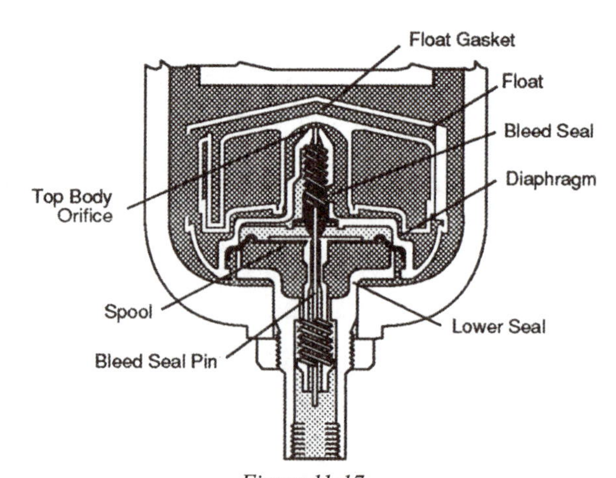

Figure 11-17

Drain in Open Position

However, as the liquid level in the bowl rises, the float moves upward, lifting its gasket off the top of the body orifice, allowing system pressure to enter above the diaphragm. This forces the diaphragm /spool downward, opening the discharge passage past the lower seal and discharging the liquid. After discharge, the float/float gasket drops and seals the top body orifice. As pressure above the diaphragm bleeds out through the clearance between the bleed seal pin and the spool nose, the spool/ diaphragm continues to rise until the spool nose seals against the bleed seal. The cycle is ready to repeat.

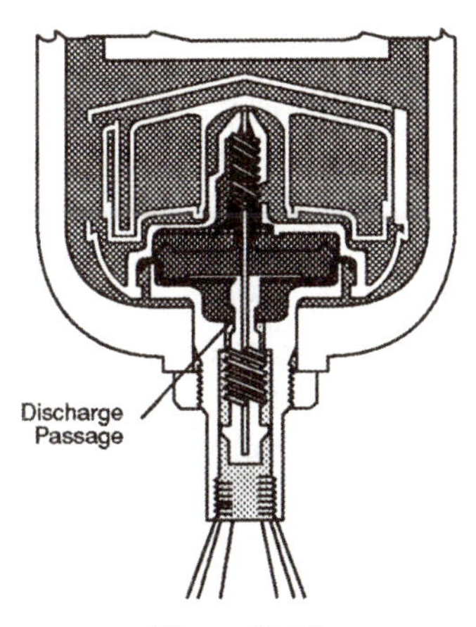

Figure 11-18

Both types of drains are found at the bottom of the filter bowl. These bowls can be made of many different materials.

Metal Bowls and Bowl Guards

Many of the filters manufactured make use of polycarbonate for bowl materials. This is generally a good material for filters and lubricators because it is transparent and tough. Properly designed, fabricated and maintained, it is suitable for use in a normal industrial work - place environments, but should not be located in areas where it could be subject to impact blows or temperatures outside of the rated range.

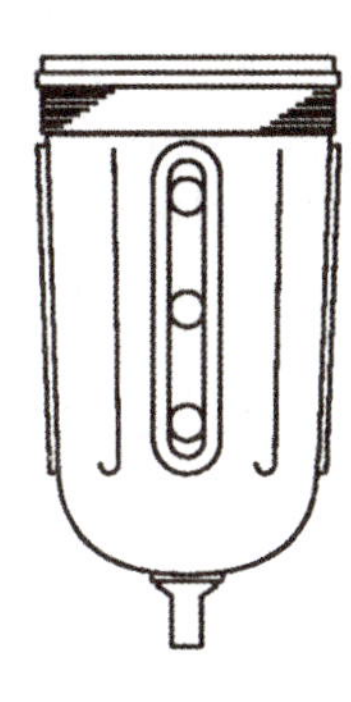

Metal Bowl

Figure 11-19 Metal bowl

As with most plastics, exposure to certain chemicals can cause damage. For example, polycarbonate bowls should not be exposed to chlorinated hydrocarbons, ketones, phosphate esters and certain alcohols. They should not be used in air systems where compressors are lubricated with fire-resistant fluids such as phosphate ester or di-ester types. Metal bowls resist the action of most solvents but should not be used where strong acids or bases are present or in salt-laden atmospheres.

For most applications suitable for polycarbonate bowls, a bowl guard is recommended. This perforated metal shield fits around the bowl exterior to protect the bowl from mechanical damage as well as to contain the bowl parts in a case of bowl rupture.

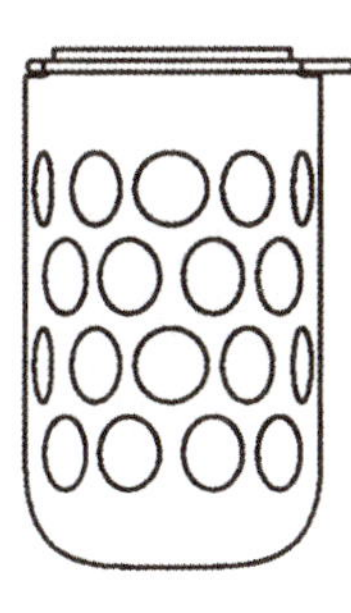

Bowl Guard

Figure 11-20 Bowl guard

Selecting a Filter

Let us now select a filter. When selecting a filter these three steps should be followed:

1. Determine the maximum flow requirements where the filter will be used. This will be in terms of cfm (dm^3/s).
2. Determine the allowable pressure drop at the needed flow.

3. Refer to the flow curves on the filter to find one that offers the particular flow rate at the predetermined pressure drop.

Remember, if an element is sized too small, the result will typically be a high initial pressure drop at the rated flow. Also, the element will "load up" quickly at the start of operations with a resulting increase in pressure drop. This can lead to possible air starvation downstream. When the pressure drop reaches 10 psi, the filter element should be changed in order to prevent its rupture. If the filter is too large, the low swirl velocity generated may result in poor contaminant and condensate separation.

Example 11-1

The XYZ Machinery Company is using a radial type piston motor which develops a maximum output of 10.0 hp (7.5 kW) at 800 rpm. Its air consumption, as read directly from catalog data, is 200 cfm (95 dm³/s). Select a filter that will work with a maximum pressure drop of 2 psi (0.14 bar). The data given for the motor is at 85 psi (6 bar).

Solution:
From motor catalog data it is evident that we need a filter that will pass 200 cfm at 85 psi, with a 2 psi pressure drop.

Filters: Performance Characteristics

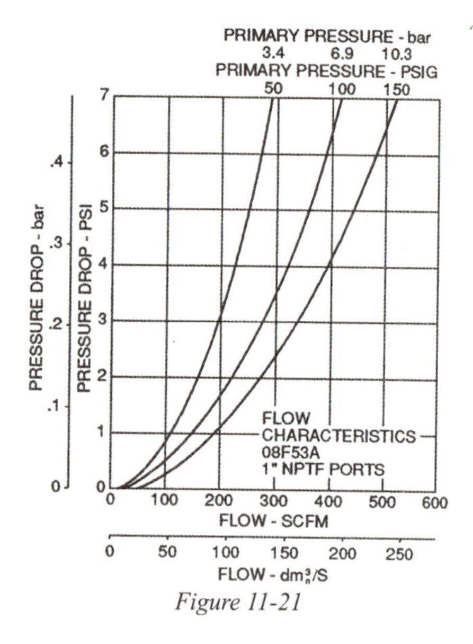

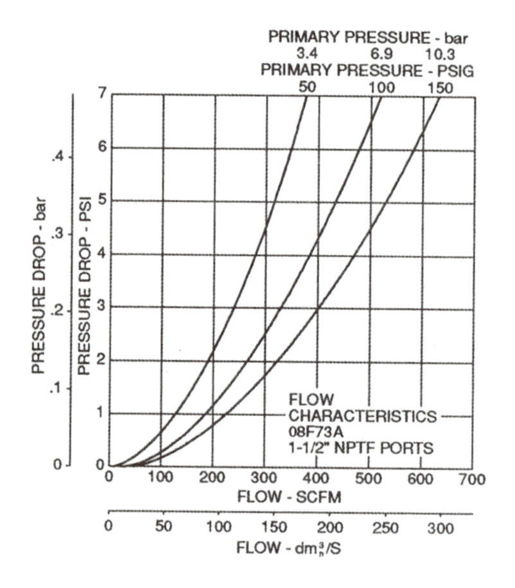

Figure 11-21

As we go through a filter catalog we come upon those two graphs showing the flow characteristics of a basic 1" port body with 1" pipe and 1½" pipe ports. Since the primary pressure of 85 psi (6 bar) is not given, it must be estimated as shown (Figure 11-21).

Filters: Performance Characteristics

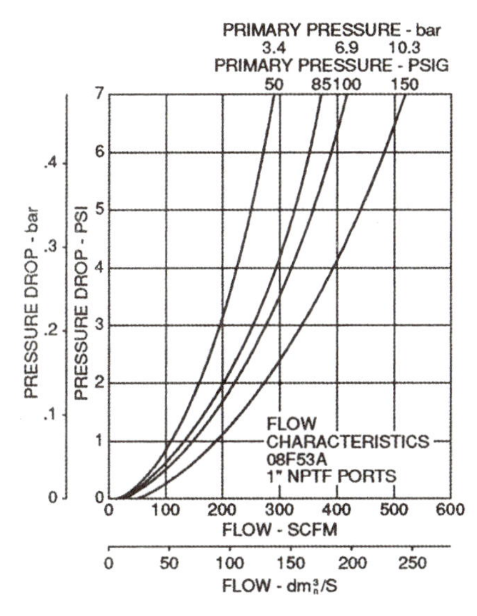

Figure 11-23

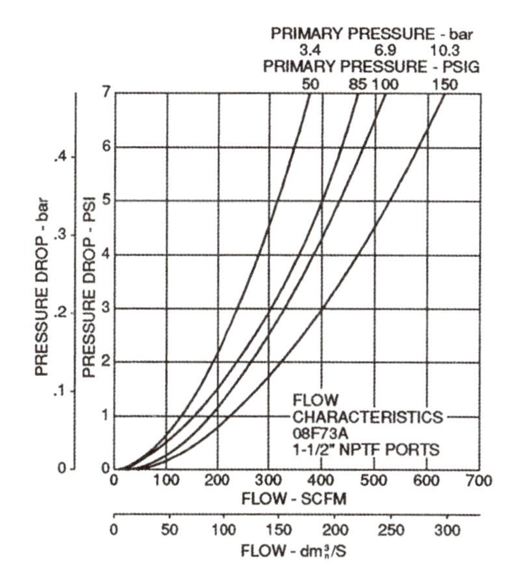

Figure 11-24

Working with the 08F53A graph, we find a 2 psi (0.14 bar) drop for 200 cfm if 1" plumbing is used. A 1.5 psi (0.1 bar) drop is evident for the 08F73A if 1½" plumbing is used. Either may be selected but the former has better water removal efficiency.

Example 11-2

A 6" (152 mm) bore, 3" (76.2 mm) rod, 10" (254 mm) stroke cylinder extends in 4 seconds and retracts in 3 seconds. The working pressure required at the cylinder is 50 psi (3.5 bar). The filter's primary pressure is 100 psi (7 bar). What filter is needed?

Solution:

The cfm must be calculated first from Chapter 6.

$$\text{cfm} = \frac{\text{Volume (in}^3\text{) x Compression Ration}}{\text{time to fill cylinder(s) x 28.8}}$$

For extension

$$\text{cfm} = \frac{(28.27 \times 10) \times \left[\frac{50 + 14.7}{14.7}\right]}{4 \times 28.8}$$

cfm = 10.8 or (5.09 dm³/s)

For retraction

$$\text{cfm} = \frac{\frac{\pi}{4}(6^2\text{-}3^2) \times 10 \times \left[\frac{50 + 14.7}{14.7}\right]}{3 \times 28.8}$$

cfm = 10.8 or (5.09 dm³/s)

The filter must be able to pass 10.8 cfm (or 5.09 dm³/s) for either direction of cylinder motion, with a primary pressure of 100 psi (7 bar).

Let us say the allowable pressure drop is 1.5 psi (0.1 bar). Looking through the catalog, we find the following graphs (Figure 11-22).

Filters: Performance Characteristics

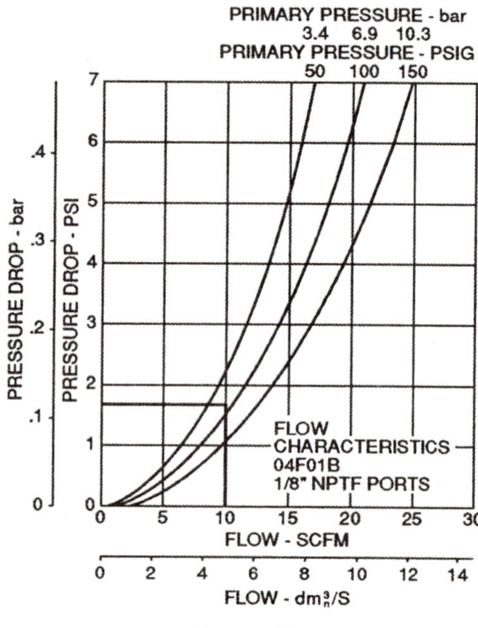

Figure 11-25

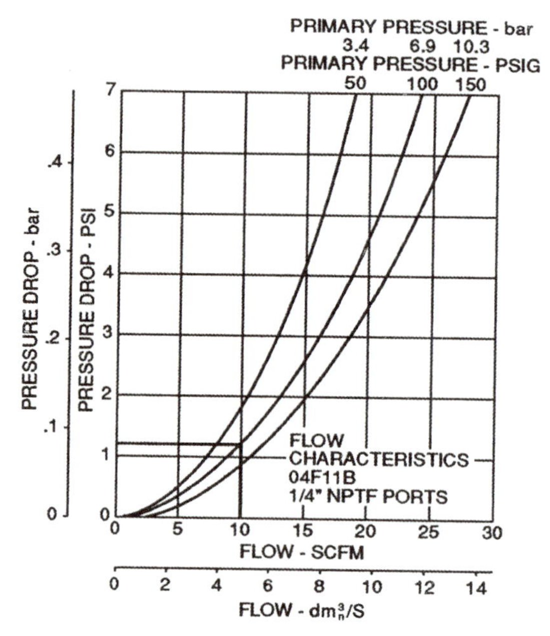

Figure 11-26

The filter selected will be the 04F11B with ¼" ports which will deliver 10.8 scfm (5 dm³/s) with 1.2 psi (0.08 bar) pressure drop.

However, it is common practice to select a larger filter to facilitate ease of piping. A 6" bore cylinder typically would have 1" ports. If this larger filter is selected then it is a must that the component's contaminant removal efficiency be examined.

Now that we have filtered the air we must add a lubricant to the system. This can be done with an inline lubricator.

Lubrication

Some forms of lubrication found in a pneumatic system are accidental. They are caused by condensed moisture and oil picked up during compression being carried downstream to the working components. This type of lubrication is very crude and usually causes more problems then it cures. Deliberate lubrication can be provided by periodic injection of oil through an extra port or grease through a fitting, by drip type oilers or by automated pressure injection of the lubricant.

Most common today is the introduction of oil droplets into the air supply by devices called air line lubricators. When injected into the flowing stream the oil in the lubricator can take on several forms - drops, spray or mist.

Pulse Type Lubricators

Drops injected into an air line are often done by a pulse type lubricator. This method injects a given minute amount of lubrication directly into the pneumatic component. Oil injection is directly proportional to cycle rate. This type may require an additional line to be run to each pneumatic component requiring lubrication.

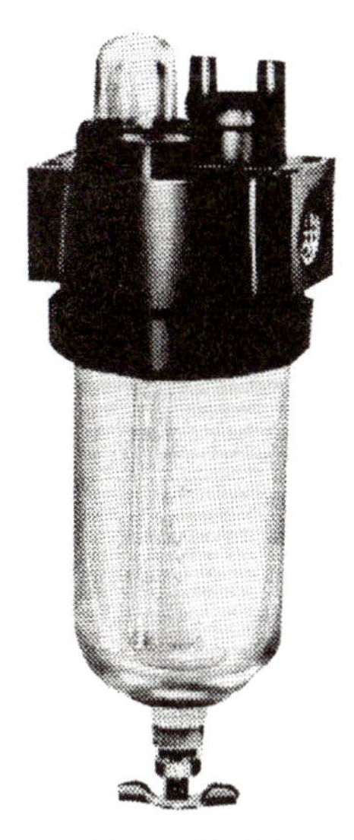

Figure 11-27

Standard Air Line Mist Lubricator

Another method is to spray the oil into the flowing air stream. This spray of oil may be obtained in an air line by applying a standard air line mist type lubricator. This type delivers oil particles ranging in size from 0.01 to 500 micrometers in diameter.

How a Standard Air Line Mist Lubricator Works

Air flowing through the unit goes through two paths. At low air flow rates, the majority of air flows through venturi section (A). The rest of the air slightly deflects and flows past the restrictor (J). The velocity of air flowing through the venturi section creates a lower pressure at the throat section (B). This lower pressure allows oil to be forced from the reservoir through the pick up tube (C), past the check ball (D), to the metering block assembly (E). This is where the oil flow rate is controlled by the metering screw (F). Rotation of the metering screw in a counterclockwise direction increases the oil flow rate; conversely, rotation in a clockwise direction decreases it. Oil then flows through the clearance between inner and outer sight domes (G), where drops are formed and drip into the throat section (B). Here it is broken into fine particles and mixed with the swirling air to be carried to the outlet where it joins air bypassing the restrictor disc (J).

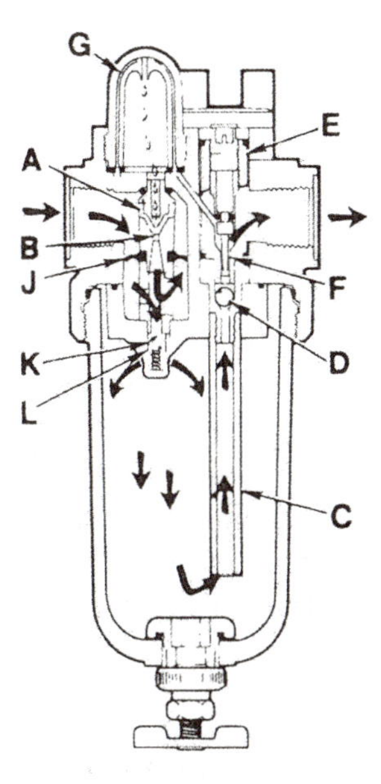

Figure 11-28

As air flow increases, the restrictor disc (J) deflects, allowing the additional air to bypass the venturi section. This also generates an increased pressure drop across the venturi and will increase the oil delivery rate proportional to the increased air flow rate. The check ball (D) assures that when there is no air flow, oil in the passageways is held in place shortening the time required to resume oil delivery when flow is re-established.

Filling the Standard Air Line Mist Lubricator

For the standard lubricator, the bowl can be filled while the lubricator is at pressure in a working system. This is made possible by the action of the check ball (K). When the fill cap is partially opened, air in the bowl escapes and pressure forces the check ball to nearly seal at (L). **Only** then may the cap be removed and oil poured into the lubricator, when the fill cap is replaced, a small amount of air bleed past the check ball (K) causes pressure to build up in the bowl allowing the system to repressurize.

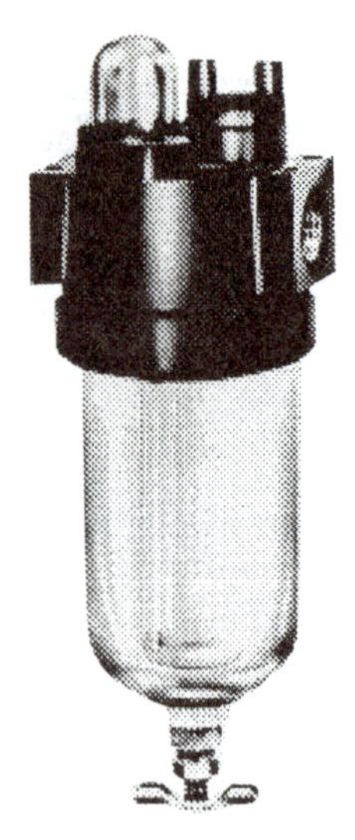

Figure 11-29

Recirculating Type Lubricator

There is another type of lubricator, the recirculating or micromist, flow in-line lubricator. This type mists the oil very finely for use downstream. The oil particles coming from the outlet range in size from 0.01 micrometers to 2 micrometers in diameter. These smaller droplets (as compared to the standard lubricator) will stay in suspension over a much greater distance allowing longer lines. Also, elbows and vertical runs will not wet (coalesce) out the droplets as quickly as the standard unit. This is because the larger droplets are returned to the bowl by recirculating the air flow into the bowl.

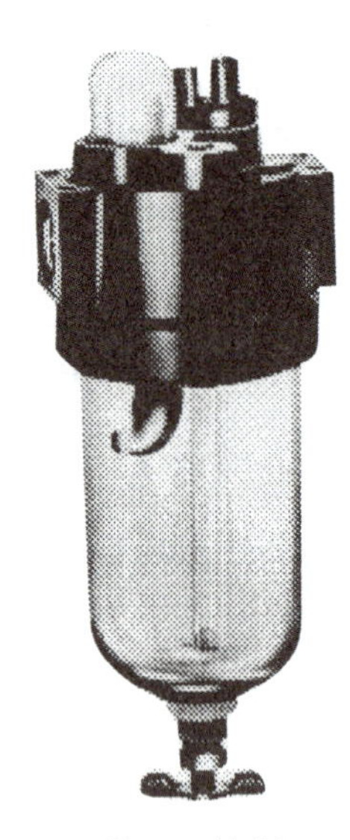

Figure 11-30

How a Recirculating Type Lubricator Works

Air flowing through the unit goes through two paths. At low air flow rates, the majority of the air flows through the venturi section (A). The rest of the air slightly deflects and flows past the restrictor disc (J). The velocity of the air flowing through the venturi section (A) creates a lower pressure which allows oil to be forced from the reservoir through the pick up tube (C) past the check ball (D), to the metering block assembly (E). It is at this point that the oil delivery rate is controlled by a metering screw (F).

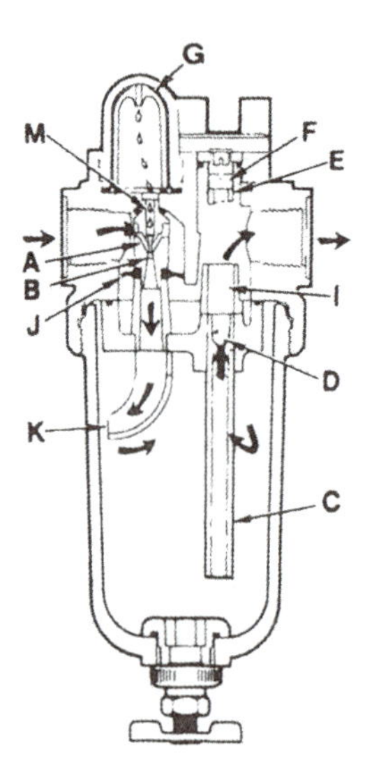

Figure 11-31

Rotation of the metering screw in the counter-clockwise direction increases oil flow rate while clockwise rotation decreases it. Oil flows through the clearance between the inner and outer sight domes (G) where drops are formed and drip into the nozzle tube (M). Here it is broken into fine particles as it expands into the low pressure venturi. From there, the atomized oil flows through the curved baffleplate (K) and is deflected against the interior wall of the reservoir. This action causes larger oil particles to coalesce and fall back into the reservoir where it can recirculate through the system. The remaining mist of fine particles (typically less than 2 micrometers) is carried through the opening (I) where it joins and mixes with air that bypassed the restrictor disc. As air flow increases, the restrictor disc deflects, allowing more of the inlet air to bypass the venturi section. This also generates an increased pressure drop across the venturi and will increase the oil delivery rate proportionate to the increased air flow rate.

The check ball (D) prevents reverse oil flow down the pick up tub when air flow stops. Thus, oil delivery can resume virtually immediately when air flow restarts.

It should be noted that this type of lubricator, as described, can **only** be filled with the supply pressure SHUT OFF. If this is unacceptable, options for refilling with auxiliary pressure supply equipment are usually obtainable (refill check valve & auto fill accessory).

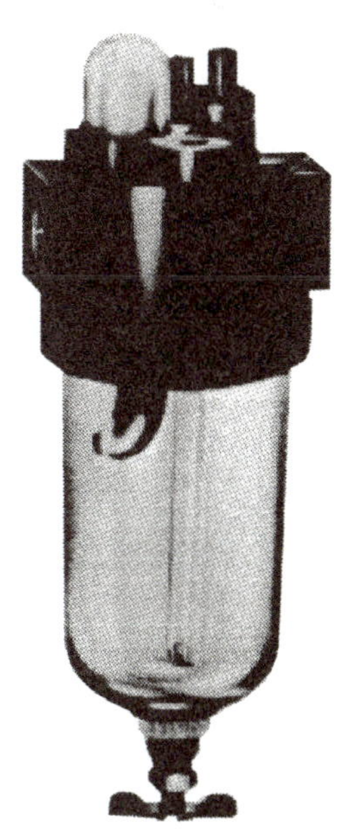

Figure 11-32

Sizing a Lubricator

The lubricator, like the filter, is sized on air flow rate. Once again, the maximum flow requirements of the system must be determined and the pressure drop (usually in the range of 1 to 5 psi [0.07 to 0.35 bar]) for various body sizes and plumbing must be determined. Then catalog data is checked and a lubricator is selected.

Example 11-3

A die grinder employs a vane motor. At maximum output the motor needs 16.3 cfm (8 dm^3/s) at 50 psi (3.5 bar). What lubricator should be used?

Solution:
The lubricator selected must be able to pass 16.3 cfm (8 dm^3/s) as a reasonable pressure drop. The pressure drop selected will be 5 psig (0.34 bar). The following graphs (Figure 11-28) are those typically found in a catalog for micro-mist lubricators.

The pressure drop for a ½" body size lubricator is 2.3 psi, 1.5 psi and 1.3 psi for ¼", ⅜", ½" ported respectively. All lubricators have a reasonable pressure drop. They all fall in the range of 1 to 5 psi (0.07 to 0.35 bar). Since the plant engineer prefers to buy a lubricator with the same size ports as the motor, the ¼" model is selected.

Lubricators: Performance Characteristics

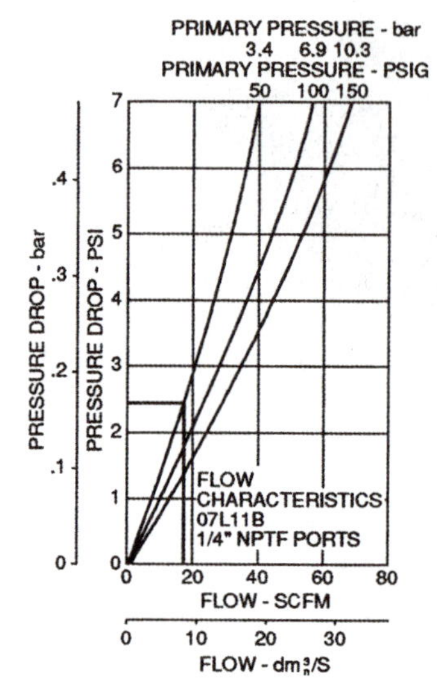

Figure 11-33

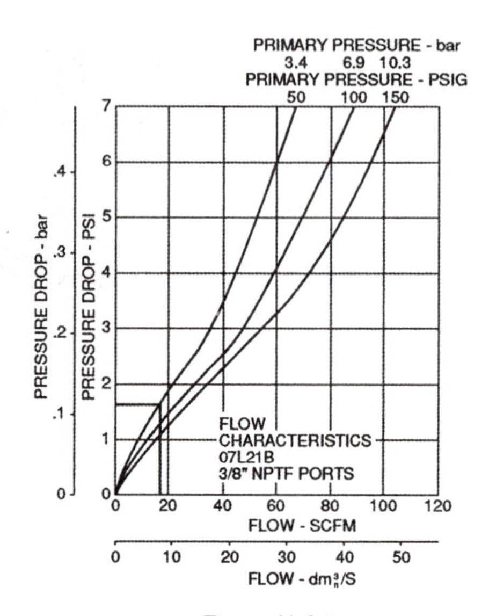

Figure 11-34

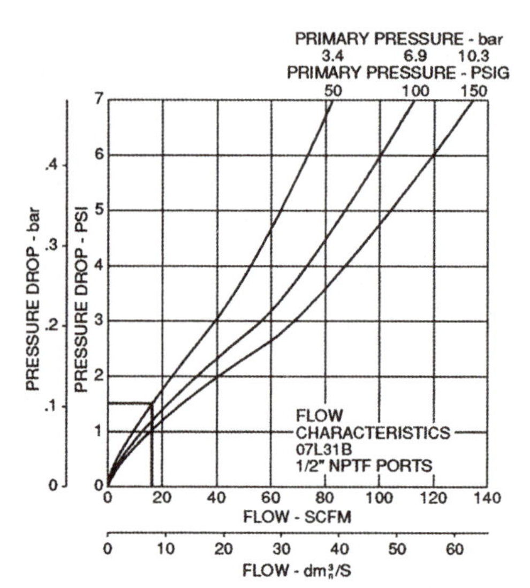

Figure 11-35

The Lubricator in a System

It should be pointed out that all the lubrication devices discussed were necessary in order that oil is carried through the circuit with the compressed air. Proper lubrication is necessary because it is needed to coat resilient seals to reduce friction, significantly extending their life. Most pneumatic cylinders and motors also require lubrication to decease friction and prevent scoring, thus decreasing operating temperature and preventing scoring. This will increase the life of the motor or cylinder.

However, just because a little oil is good doesn't mean a lot of oil is better. Many times, lubricators are adjusted so that too much oil is injected into the stream. Puddles of oil develop in the piping and cause problems. Also, air exhausting with large amounts of entrained oil may cause enough oil to become airborne in the plant that the air in working areas will not meet federal standards. Remember the oil put into the air system must be taken out before the air is returned to the atmosphere. The OSHA regulation states that air must not contain more than 5 mg of oil mist particles per cubic meter of air. This is equal to approximately an ounce (wt.) of oil in 200,000 cubic feet which is equivalent to a building 10 x 100 x 200 feet in size. If a plant uses dozens of lubricators it may take a very long time for the oil to settle out. The limit of 5 mg per cubic meter may quickly be exceeded when the plant is operating. This OSHA requirement makes it necessary that you use some kind of device to reclassify the oil before the air carrying it is exhausted into the atmosphere. Typical inline air filters do a relatively good job of reclassifying oil. But if almost total reclassification is necessary, a coalescing type filter can be employed. It is common to place all of these units together to form an FRL.

FRL Units

FRL stands for filter-regulator-lubricator. An FRL unit combines the three components into a pre-piped package for easy installation. Branches of a pneumatic system are generally equipped with FRLs so that individual actuators and work stations receive filtered, regulated and lubricated air meeting their specific requirements.

Figure 11-36

Filters and Lubricators Must be Maintained

A pneumatic system may be equipped with the best filters and lubricators available and they may be positioned in the system where they do the most good, but if filters are not drained and changed at reasonable intervals for such services, or lubricator reservoirs re-filled, the money spent for their use may have been wasted (Figure 11-30).

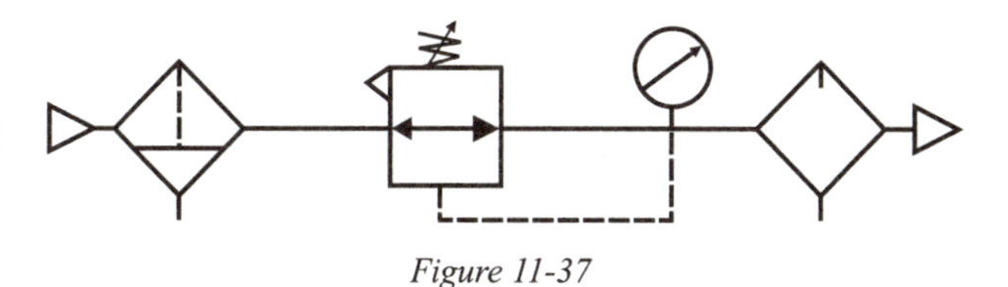

Figure 11-37

FRL Design Considerations

The following is a set of definite rules that should be followed when designing systems protected with FRLs.

1. Make sure they are sized for the maximum flow rate in the portion of the circuit they service.
2. Placement of these devices should be as close as practical to the component being serviced.
3. Their placement should also be in an area of easy accessibility. This is especially true for units requiring regular maintenance, such as filling, adjusting or cleaning.
4. Install a shutoff valve (possibly a lockout type, shutoff exhaust) ahead of the unit to facilitate maintenance of the unit.
5. Whenever possible, locate lubricators at an elevation higher than or equal to the points being lubricated.

Figure 11-38

Lesson Review

In this lesson dealing with air preparation, we have seen that:

- Pneumatic systems and components require compressed air free of contamination for dependable operation.
- Dirt in an industrial pneumatic system is any solid particle in an air stream.
- Dirt in a pneumatic system does harm by causing resilient seals and closely fitted moving surfaces to wear.
- Dirt wedges into clearances between moving parts of components.
- Dirt in a pneumatic system is pollution which is measured on the micrometer scale.
- Industrial air is a carrier of dust and water vapor.
- A compressor intake filter is the first line of defense against industrial air.
- Entrained water and large dirt particles are removed by a swirling action set up within the filter bowl.
- The function of a mechanical filter element in a pneumatic system is to remove dirt particles from a compressed air stream by forcing flow to pass through a porous material.
- Filter elements used in air line filters are divided into depth and edge types.
- Since it has no single consistent hole or pore size, an air line filter element is given a nominal rating which is designed to indicate an element's average pore size.
- No guarantee is made by an air line filter nominal ratings as to what size particles will be removed from a compressed air stream.
- Instead of discharging air from a compressor outlet into an air receiver for storage, the air is passed through an aftercooler.
- Besides being the point where air cools, an aftercooler is also the place where some dirt and oil vapor fall out of suspension and a good portion of entrained water vapor condenses.
- If the demands of a system require that very dry air be made available, compressed air can be passed through a refrigerant or chemical type air dryer.
- Moisture can affect the operation of a pneumatic system in several ways.
- After compressed air passes through an air line filter, it is clean and relatively free of entrained water, but is usually required to be lubricated.
- Lubrication of compressed air may be necessary to provide seal lubrication, prevent sticking of moving parts and control wear.
- An FRL unit combines filter, regulator and lubricator into a pre-piped package for ease in installation.
- If filters and lubricators are not serviced when required, the money spent for their use has been wasted.

Exercise
Air Preparation
50 Points

Instructions: Fill in the blanks with a word from the list at the end of the exercise. Words may be used more than once.

1. Entrained water and large dirt are removed by an air line filter by means of a(n) ______________________ action set up within the filter bowl.
2. A(n) ______________________ is a component in which air cools, some dirt and oil vapor fall out of suspension, and a good portion of entrained water vapor condenses.
3. A nominal rating indicates an air line filter's ________________ pore size.
4. ____________________ in a pneumatic system can wash away lubricant from components, resulting in faulty operation, corrosion and wear.
5. As air leaves a compressor, it has more potential energy, but it is also______________ and contains water vapor.
6. To prevent sticking of moving parts and to control wear, compressed air requires ____________________ .
7. Air line filter elements are given a _______________ rating since they have no one, consistent hole or pore size.
8. The more contaminated the air of a pneumatic system, the less __________________ that system will be.
9. Water vapor and _________________ are two undesirable elements carried by industrial air.
10. In a pneumatic system, dirt particles are measured using the_________________ scale.

absolute	depth	intercooler	porous bronze
aftercooler	dust	largest	quiet zone
average	edge	lubrication	rain
baffle	FRL	maintenance	refrigerant
chemical	grain of salt	micrometer	salt
compressor	hot	mist chamber	swirling
condensation	industrial air	moisture	vane
dependable	intake filter	nominal	wedges

Index

A

absolute
- pressure scale . . . 10, 12, 17
- temperature . . . 17

absorption . . . 50
accuracy of a flow control . . . 105
actuators . . . 7, 92, 93
adjustable orifice . . . 104
aftercooler . . . 48
air
- compression . . . 17
- cylinder . . . 70
- expansion . . . 18
- fuses . . . 126
- leaks . . . 55
- line filter . . . 131
- motors . . . 76
- selection . . . 76, 77
- oil tanks . . . 124
- to-air booster circuits . . . 125
- to-oil booster circuits . . . 125

altitude . . . 41
ambient . . . 17
atmospheric pressure . . . 9
automatic drain . . . 134
axial compressor . . . 37

B

baffle plate . . . 131, 132
ball valve . . . 104
barometer . . . 10
bias spring . . . 11
blocked center . . . 85, 86
boosters . . . 123, 124
- circuits . . . 125

Bourdon tube gage . . . 12
bowls . . . 132, 135
- guards . . . 135

C

cfm . . . 21
C_v . . . 95-99
center condition . . . 85
- blocked . . . 85, 86
- exhaust . . . 85
- pressure . . . 85

centrifugal compressor . . . 38
check valve . . . 59
circle, area of . . . 6
circuits
- air-to-air booster . . . 125
- air-to-oil booster . . . 125
- differential pressure . . . 121
- dual pressure . . . 121
- four-way valve in . . . 83
- quick exhaust valve in . . . 109
- regulator in . . . 120
- sequence valve in . . . 115
- three-way valve in . . . 82, 83

compressed air, industrial applications . . . 3
compression factor . . . 96
- heat of . . . 20
- of air . . . 16

compressor . . . 28
- axial . . . 39
- centrifugal . . . 38
- displacement . . . 47
- dynamic . . . 47
- helical . . . 37
- installation . . . 45
- lobed-rotor . . . 38
- multi-stage . . . 39
- radial . . . 39
- reciprocating piston . . . 37
- selection . . . 44
- single screw . . . 38
- two-stage piston . . . 40
- vane . . . 37

contaminants 129
built-in 129
entrained dirt 130
generated 129, 130
hard dirt 130
ingested 130
soft dirt 130
critical velocity 22
cushions 66
cylinders 7, 57
buckling 64
mechanical motions 59
mounting styles 59
seals 58
sizing 60-72
stroke adjustors 58
volume of 9

D

dashpot, hydraulic 106
deflector plate 131
depth filter elements 132
detents 84, 85
diaphragm regulator 118
differential pressure circuit 121
directional control valves 30, 81-101
displacement compressor 47
double-acting cylinder 30
drain
automatic 134
filter 134
open position 135
dual pressure circuit 121
dynamic compressor 47

E

edge filter elements 132
excess flow valves 126
exhaust center 85
externally piloted valves 93, 94

F

FRL units 143
filter
air line 131
element 131, 132
depth type 132
edge type 132
pore size 132
oil removal 133
rating 133
selecting 135-138
fixed orifice 104
flow
coefficient 95-99
control valves 31, 103-109
rate 21, 70
rating 95-99
fluid direction 21
force 8, 60
four-way directional valve 83
in a circuit 83
five-ported 83
four-ported 83
free air 22
friction 21

G

gage pressure scale 9
gaseous fluid 5
gases 15
pressure 16, 17
temperature 16, 17
globe valve 104

H

heat
energy 16
of compression 20
helical compressor 37
hydraulic dashpot 106

I

Ideal gas law 17
industrial applications of compressed air 3
intensifiers 8, 123, 124
internally piloted valves 93

K

kinetic energy 5, 18

L

lapped spool valves 89, 90
levers 92
limit of visibility 131
liquid molecules 16
lobed-rotor compressor 38
lubrication 138
lubricators
FRL units 143
pulse type 139
recirculating 140
sizing 141
standard air line mist 139, 140

M

manually actuated valves 92
manually operated valves 92
mechanical force 7
mercury 10
meter-out control 105
micrometer scale 131
molecular energy 15
motors 74
piston 75
turbine 76
vane 75
mufflers 11
multi-stage compressor 39

N

needle valve 31, 104
noise 42, 110
nominal rating 133
normally closed valves 84
normally open valves 84

O

oil flooded compressor 38
oil removal filter 133
orifice
adjustable 104
fixed 104
size 103
output control 41
overcompression 49

P

psi 10
psia 10
psig 10
packed bore valve 88, 89
packed spool valve 87, 88
Pascal's Law 6
pedals 92
percussive 72
pilot actuators 93
solenoid controlled 94
pilot controlled regulators 118, 119
piping systems 53
dead end or grid 54
installation 54, 55
loop 54
unit or decentralized 54
piston motor 75
plunge pressure gage 11
plungers 92
pneumatic
symbols 32
system design 23
system inefficiency 20
tools 72
transmission of energy 18
valve 22
potential energy 5, 19
pore size filter elements 132
port pipe size 95
positive displacement compressor 19, 20
power positions of a valve 85
pressure
applying 6
atmospheric 9
center 85, 93
differential gage 46
gages 11
of a gas 16
regulator 29, 116
scales 9
switch 28
pulse lubricator 139
pushbuttons 92

Q

quick exhaust valves 109-111
quiet zone 132

R

radial compressor 39
receiver tank 51
reciprocating piston compressor 36
recirculating lubricator. 140
refrigeration 50
regulators
    disphragm 118
    FRL units 143
    in a circuit 120
    pilot controlled 118, 119
    pressure 116
    sizing. 119
venting 116, 117
response time of directional valves 98, 99
rotary motors 74

S

safety relief valve 28
sandwich flow controls 105
seals. 58
selecting a filter. 135-138
sequence valve. 115
shear action valves 86
    lapped spool. 89, 90
    packed bore 88, 89
    packed spool 87, 88, 89
    sliding plate 86, 87
shroud. 132
shuttle valves. 111
silencers 111
single screw compressor 38
sizing
    compressors. 60-72
    lubricators 141
    regulators 119
    valves 95-98
sliding plate valve. 86, 87
solenoids 92
    controlled pilot operation 94
    directing operating 92, 93
solids. 16
speed control
    double-acting cylinder. 106
    multiple 107
spring offset valve. 84
spring returned valve 84
standard air 22
standard airline mist lubricator 139, 140
stop tube. 62, 63
stroke adjustors 58
sub-base mounted valves 94, 95
subsonic flow. 114

T

temperature absolute. 17
temperature of a gas 16
three-position valves 84, 85, 92
three-way directional valve 82, 83
    in a circuit 82
    paired 82
torque 76
turbine motor 76
two-position valves 84, 92
two-stage piston compressor 40
two-way directional valve. 81

V

vacuum. 11
    gage. 12
    pressure scale. 10
valves, directional 81
    actuators. 92, 93
    C_v. 95
    center condition of 85, 86
    detented 84, 85
    externally piloted 93, 94
    flow rating 95
    four-way. 83, 85
    internally piloted. 93
    manually actuated. 92
    manually operated 92
    normally opened. 84
    pilot operated 94
    poppet valve. 91
    power positions. 85
    response time. 98, 99

shear action 86
lapped spool 89, 90
packed bore. 88, 89
packed spool 87, 88, 89
sliding plate 86, 87
sizing. 94-98
solenoid 92, 93, 94
spring offset 84
spring returned 84
sub-base mounted. 94, 95
three-position 84, 85, 92
three-way 82, 84
two-position 84, 92
two-way 81
valves, excess flow. 126
valves, flow control 103
accuracy 105
ball valve. 104
globe valve 104
meter-out control 105
needle valve 104
orifice size 103, 104
sandwich configuration 105
valves, quick exhaust 109
as a shuttle 111
in a circuit 111
valves, sequence 115
in a circuit 115
vane compressor. 37
vane motors. 75
velocity, critical. 22
velocity of a fluid 22
ventilation 43
venting regulator 116, 117
visibility, limit of 131
volume of a cylinder 9

W

weight velocity 68
working energy. 27